The Cosmic
Mind-Boggling
Book

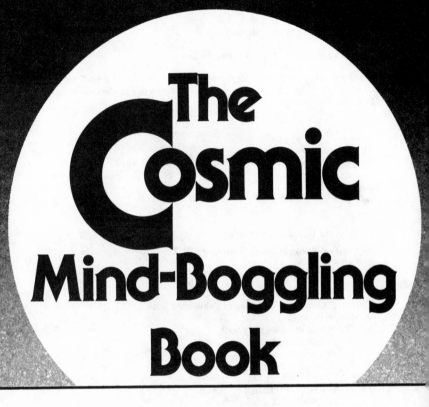

The Cosmic Mind-Boggling Book

Neil McAleer

Foreword by Robert Jastrow

WARNER BOOKS

A Warner Communications Company

All rights reserved.
Warner Books, Inc., 666 Fifth Avenue, New York, N.Y. 10103

Ⓦ A Warner Communications Company

Printed in the United States of America

First Warner Books Printing: September, 1982
10 9 8 7 6 5
Book design by H. Roberts Design
Cover design by Gene Light
Library of Congress Cataloging in Publication Data
McAleer, Neil, 1942–
    The cosmic mind-boggling book.
    Bibliography: p.
    Includes index.
    1. Astronomy—Popular works.   I. Title.
QB44.2.M4   1982   523   81-14677
ISBN 0-446-39046-1 (U.S.A.)   AACR2
    0-446-39047-X (Canada)

To Barbara and Jim
with love

CONTENTS

FOREWORD

Dramatic discoveries have transformed astronomy during the last ten years. No longer is the Universe conceived of as a cold and empty, unchanging place; today we know it to be richly populated by exotic objects, lashed by savage forces, and pregnant with surprises. Strange quasars light up the dark corners of the Universe; massive galaxies race across the sky; titanic explosions of unknown origin occur in the depths of space; and tentative evidence has been uncovered for the black hole—the most bizarre object ever conceived by the scientific mind.

These new discoveries have raised cosmic consciousness to a new level. Words like Cosmos, light-year, and galaxy are found everywhere. The mind grows accustomed to their constant repetition; after a time, it mistakes familiarity for understanding; yet the simplicity of these terms is deceptive. Who can truly comprehend a Universe in which the natural dimensions of time and space are a billion years and a trillion miles? Who can grasp the meaning of the brief span of human existence in a world in which suns and earths disappear like autumn leaves falling to the ground?

It has been the inspiration of Neil McAleer to explain these radical notions by breaking them down into many fascinating nuggets of information, each one digestible in itself, and each a small revelation. The reader browses through the facts with wonder and delight. Then—and this is the great value of Neil McAleer's book—the reader stands back, and suddenly he sees the larger picture emerging out of the details; for the first time, he understands the grandeur of the Universe and the smallness of humankind's place in it.

The humble positions of the Earth and humankind in the Universe—these thoughts are the essence of the Copernican Revolution. Before the time of Copernicus, it was obvious to anyone who followed the motion of the Sun day after day, and the motions of the Moon and the stars night after night, that the Earth is the center of the Universe and that the heavenly bodies revolve about it daily, paying homage to the abode of man. Every day the Sun moved across the vault of the heavens; every night the Moon and stars traveled in their stately procession across the sky. *Not so,* said Copernicus, 500 years ago. *The Earth moves,* he said. *The Earth moves around the Sun every year, and turns on its axis every day.* No longer is the Earth at the center of the Universe; now it is the Sun that "sits in the midst of all enthroned" and is "rightly called the Lamp, the Mind, the Ruler of the Universe, gathering his children the Planets, which circle around him."

Martin Luther called Copernicus the "fool" for contradicting the Bible. Yet the Copernican theory took root in men's minds. There was an air of freshness about it; it opened the door on new ideas that went far beyond the science of astronomy. Humankind saw the implications in the sun-centered Universe. Consider the following facts, they said: The five planets revolve around the Sun; the Earth revolves around the Sun; since the Earth and the planets behave in the same way, they must be the same kinds of objects. This was a startling conclusion. Prior to that time, astronomers had believed that the planets were hard, polished spheres of jewel, perfect and unchanging, while the Earth was made of mud, rock, and water. If the Earth and the planets are similar objects, the planets may also be made of mud, rock, and water. One unsettling thought led to another: If the planets are made of the same substance as the Earth, perhaps they bear life; perhaps they have people . . .

Those were radical ideas. They presaged the latest developments in twentieth-century science, which seek to unite life on the Earth and life in the Cosmos. The astronomy of Copernicus was the first step in the Copernican Revolution; Copernicus removed the Earth from the center of the Universe and put the Sun in its place. Others took the second step; they removed the Sun from the center of the Universe and put nothing in its place. *There is no center,* they said; the Universe is infinite, and contains an infinite number of stars. Each star is a sun like ours, each has a family of planets.

These thoughts led finally to the modern picture of a Universe populated by innumerable suns, innumerable earths, and, perhaps, innumerable forms of life. In that image we see the ultimate truth: Man stands at the summit of creation on the earth, but in the cosmic order his position is humble. No revelation more striking has ever come from the scientific mind.

Robert Jastrow

PREFACE

There is a realm where human eyes cannot see, where hands cannot touch, where all human senses are overwhelmed and life spans shrink to insignificance. This realm is almost everywhere, all around us. It is the rest of the Universe—everything there ever was, is, or will be; everything our senses cannot know directly.

For the first time in 2 million years of human history, we are gaining the capacity to translate, describe, and understand the Universe: its beginning, middle, and end and its cast of cosmic characters: dwarf, giant, and supergiant stars; pulsing and exploding stars; black holes; myriad-shaped galaxies, some calm, some in violent upheaval; and the ancient quasars, the oldest cosmic objects known, whose radiations have traveled for billions of years, from a time when the Universe was young.

The Universe has been brought "down to Earth" only in the last few decades, through a combination of human diligence, brainpower, and extrasensory technology, by some of the great minds of the twentieth century. In just 15 years (1917–32), the size of the known Universe expanded 1 trillion times.* During the great explorations of Magellan and other navigators of the late fifteenth and early sixteenth centuries, the known area of planet Earth increased tenfold. The Universe is expanding; so is the human mind, which has learned to extend the senses. We know more than ever before, but the wonder remains, and it still can overwhelm us.

The physical reality of our senses—what we see, touch, smell, hear, and taste every day—represents an infinitesimal fraction of the physical reality of the Universe. If the energy spectrum were a yardstick (36 inches; 91 centimeters), then what we see with our eyes in the small visible range would be less than a half inch (about 1.3 centimeters). But even within this limited spectrum range, the human eye quickly reaches its limits and the brain must take over. The 200-inch (508-centimeter) Hale reflector at Mount Palomar can collect 1 million times more light than the human eye. If humankind survives and the brain continues to expand its power and ability to solve problems, the future of our species should be a long one. There is time for about 167 million generations before our Sun dies. It took only about 66,000 generations to get us from the trees to the Moon.

How can the tremendous numbers with tens of zeros and the high-math equations of astronomers and physicists, both of which represent aspects of the Universe, be understood by most of us? By bringing the Universe down to Earth. Take cosmic distances, for instance. If you could drive to the Sun at

---

*"Billion" means 1,000 million (10^9) and "trillion" means 1,000 billion (10^{12}) throughout the book. The British "billion" that refers to 10^{12}, the American trillion, is not used.

55 miles (88 kilometers) per hour, it would take 193 years to get there. If your destination were the next closest star, the Alpha Centauri system (4.3 light-years away), it would take 52 million years at the same speed. A heavier foot on the accelerator and a speed of 100 miles (161 kilometers) per hour would not help much in covering the vast distances of our Galaxy. At this speed it would take 201 billion years to travel from the Sun to the center of our Galaxy—a distance of some 30,000 light-years, which represents less than a third of our Galaxy's diameter. If the diameter of our solar system were scaled to 1 inch (2.54 centimeters), then our Galaxy's diameter would be about 100,000 miles (161,000 kilometers). But there are an estimated 100 billion galaxies, and our Galaxy's mass represents only about one trillionth of the mass of the Universe.

Yet the Universe is everything—more than distance, speed, and time. A faraway quasar can equal 300 billion suns in light energy alone. But our Sun regains stature when we know that just 1 second of its total energy equals 13 million times the average annual electricity consumption of the United States. The ultimate in energy, however, occurred during the Big Bang birth of the Universe. One second after Big Bang zero, the radiation was 10 billion degrees Celsius (18.3 billion degrees Fahrenheit). Just a pinhead amount of this early Universe would give off 18 times the entire energy output of the Sun! This blazing pinhead could replace our Sun and provide the Earth with an equal amount of energy, even if it were as far away as the planet Jupiter, which is over 5 times farther from the Sun than the Earth.

This book is intended to bring the Universe down to Earth without sophisticated telescopes or complex equations, to use what our senses are able to perceive here on Earth to help us understand a Cosmos whose immensity, intensity, and infinite permutations boggle the mind.

ACKNOWLEDGMENTS

This book owes its largest debt to my consultant, T. A. Heppenheimer of the Center for Space Studies. His calculations and ideas are to be found throughout the text, and it would have been less a book without his substantial contribution. Any errors of fact or interpretation that remain are mine, of course, not his. Special thanks also go to my editor, Fredda Isaacson, who believed in the project and gave me encouragement as the work progressed.

For information and photographs, I wish to thank: Bill O'Donnell and Les Gaver at NASA Headquarters; authors Patrick Moore and Robert M. Powers; Jim Christy of the U.S. Naval Observatory; Frances B. Waranius, Ron Weber, and Mary Ann Hager of the Lunar and Planetary Institute; Audrey Likely, Amy Weiner, Dawn Dublin, and Joyce Goodwin of the American Institute of Physics; Arlene Walsh, Smithsonian Institution Astrophysical Observatory; David F. Malin, Anglo-Australian Telescope; Jack O. Burns, University of New Mexico; Sheelagh Grew, Armagh Observatory; R. G. Strom, Netherlands Foundation for Radio Astronomy; Bart J. Bok, Steward Observatory, University of Arizona; Bob Young of Science Communicators, Inc; and the many professional astronomers who generously gave photographs of their work and whose names appear in the photo credit lines.

At the foundation of this book are thousands of astronomers and physicists, most of them unnamed, whose dedicated research has brought forth an explosion of astronomical knowledge in the twentieth century. Their work is highly technical and specialized. I have done my utmost to represent their work fairly, but some distortions are no doubt generated when an author attempts to simplify the complex. Many of the astronomical values in the book are presented as definite, but the fact is that different researchers often derive different values, and astronomical measurements in general have large plus-or-minus error factors, especially for distant objects. Whenever possible, I have tried to give a consensus value.

I thank my three typists, Cathy Smith, Susan Huntsberger, and Gail O'Brien, for coming through for me at deadline time. And finally, I express deep gratitude to my wife, Connie, who so often listened to me at the end of my speechless working days.

THE SUN:
THE MIGHTY
YELLOW DWARF

☆　☆　☆　☆　☆

SOLAR BASICS

The Birth of the Sun

From clouds and their condensations come storms. The Sun was born from an interstellar cloud of gaseous matter in one of the spiral arms of our Galaxy about 5 billion years ago, and it can be considered a nuclear storm that goes on raging at a safe distance from the Earth. Humankind receives the life-giving mists (energy in the form of heat and light) that result from this faraway storm that began as a calm dark cloud of matter. It is believed that the shock waves from an exploding star, a supernova, stirred up the gases and began the gravitational collapse that eventually condensed into the rotating, disk-shaped solar nebula from which Sun and planets eventually formed. The solar nebula condensed further, and the increasing gravity at its center heated up the influx of gases. These central regions became hotter and hotter until, after perhaps 14 million years, the temperature reached the point at which nuclear processes were triggered—conversion of hydrogen into helium, which, in turn, releases energy—that have continued for the last 5 billion years. The entire evolutionary process of the Sun's birth probably took no more than 100 million years, about one one-hundredth the estimated age of the Galaxy—just another tick of the cosmic clock.

The Sun's Dim Past

The Sun was 35 to 50 percent dimmer in its infancy (4.5 billion years ago) than it is today. This is because of the increasing buildup of helium in its core, the helium created from the fusion of hydrogen. Assuming a present-day atmosphere, this dimmer Sun at noontime on a clear day would have looked like today's late-afternoon Sun or like the light during a partial solar eclipse. Several billion years from now (somewhat over 4), when the Sun nears its red-giant phase, it will be 2½ times brighter than today. But the Earth will be dark, darker than during the worst thunderstorms, because the hotter Sun will boil all the oceans away, clouding the Earth with thick blankets of steam.

Our Stellar Heavyweight

The Sun weighs 2,200,000,000,000,000,000,000,000,000 tons—as much as 332,270 planets like Earth. Expressed another way, the Sun weighs about 2.2 billion billion billion tons. Even Atlas could not take lightly the task of holding up the Sun.

The Unchallenged Dominance

The Sun contains 99.86 percent of all the substance in the solar system. The Earth contains only 1/332,000 of the Sun's mass, and on its surface live over 4.2 billion human beings. To compare the mass of a human being to the entire solar system is like comparing the mass of the lightest of atoms to a human being.

Light Weight

Sunlight has weight: that is to say, it exerts a pressure on anything that obstructs it—including Earth and the other planets. A square mile of sunlight, if it could be held in one's hand, would weigh 3 pounds. All the sunlight falling on the Earth's surface weighs 87,700 tons, about the weight of a large ocean liner. This, of course, has negligible effect, since the Earth weighs 6.6 billion trillion tons.

How Bright the Sun?

Just 1 square inch of the Sun's surface shines with the intensity of 300,000 candles. If we wanted to manufacture enough candles to equal the Sun's total brightness, the amount of tallow needed would be 10 times larger than the mass of the Earth, and the candles, bunched together, would cover a birthday cake whose circumference was as large as the orbit of the Earth around the Sun—almost 600 million miles.

Outgrowing Each Size

The Sun is the largest object in 25 trillion miles of nearby galactic space, at which distance we find the next nearest star, Alpha Centauri. The Sun's diameter of 865,000 miles is 109 times that of Earth's and 10 times that of Jupiter's. If the Sun were represented by a yellow beach ball with a 2-foot diameter, the Earth would be the size of a pea 215 feet away and

Jupiter would be the size of a large orange 1,056 feet away. Over 1 million Earths would fit inside the Sun, since it is 1 million times larger. If the Earth weighed 1 pound, the Sun would weigh more than 330,000 pounds. The Moon's entire orbit around the Earth could fit inside the Sun.

A solar-powered car that was *very* well insulated would take almost two years to travel around the Sun's equator at a save-energy speed limit.

A Famous Distance

The single best-known fact about the Sun—taught to every student in the early grades—is the Sun's mean distance from the Earth, which rounds off to 93 million miles. Since it is, however, the most accurately measured quantity in astronomy, the exact measurement should be given: 92,750,679.4 miles (149,597,870 kilometers). This Sun–Earth distance is referred to by astronomers as the *astronomical unit,* abbreviated AU. A person traveling this distance to the Sun would need:

21 years if flying at 500 mph (804 kph)
193 years if driving at 55 mph (88 kph)
2,123 years if walking at 5 mph (8 kph)

It would take more than 18,000 round trips between New York and San Francisco to equal this famous distance.

Old Man Sun: How Old?

Only in recent history has the question of the Sun's age been taken seriously. The ancients considered it ageless and immutable, but contemporary science has taught us just the opposite—the Sun was born and it will die. Solar scientists put the Sun's age at 4.65 billion years old, at least, and the rounded-off figure is generally given as 5 billion years. This age estimate is the result of this century's revolution in nuclear physics and of terrestrial, lunar, and meteorite rock datings. Lord Kelvin, a prominent nineteenth-century scientist, calculated that the Sun's age could not be over 100 million years; he was far from the mark only because of his ignorance of nuclear physics. His 100-million-year age would have been correct for certain stars—for those larger than the Sun (the rule appears to be that the larger they are, the faster they burn) and for newly born stars (for example, those in the Pleiades cluster). Life spans for stars vary as much as they do for animal life—just a different scale. There are the stellar butterflies and the stellar elephants; the range is estimated to be from 1 million to 100 billion years. Our Sun seems to be average, with an estimated life span of about 10 billion years. The approximate 5-billion-year age of the Sun is 2,500 times longer than *Homo sapiens'* history on Earth.

What Is the Sun Made Of?

The Sun is an intensely hot sphere of gas, composed of the same substances as the Earth and, in fact, the Universe as a whole, which seems to serve up a universal cosmic recipe. About two thirds of the known elements have been identified in the Sun, and these are all light, since they have been found on the surface, not in the interior. Hydrogen—the dominant element in the Universe—makes up 78.4 percent of the Sun's mass, and helium accounts for 19.8 percent. The heavy elements found in the planets, which include iron, nickel, and silicon, amount to less than 2 percent of the Sun. Neon amounts to 0.2 of the Sun's mass, enough to generate a neon sign 667 times larger than the Earth. If the Sun's neon were put into standard neon tubing, the length of this tubing would be 100 million times the diameter of the known Universe. If this neon tubing were closely packed, row upon row, and made into a square neon sign 3.66 trillion miles (5.9 trillion kilometers) on a side, and then placed as far away as Barnard's star (6.1 light-years), a flashing message, "Hello, Earth," would be clearly visible, just like a sign on Broadway or Piccadilly, and would appear 12 times larger than the Moon.

Heavy Stuff, Light Stuff

The Sun, because of its gigantic size, has a great range of densities, from the least dense corona that surrounds it like a bright halo to the extremely dense core where the intense nuclear reactions take place. The Sun's average density is about 1.4 times that of water (the Earth's is 5.5), which would make a piece of it weigh about as much as an equally sized lump of soft coal. But when particular layers of the Sun are considered, comparisons become more difficult. The solar gas of the Sun's surface, the photosphere, is so thin that it would be considered a vacuum on Earth. In the Sun's core, however, which is estimated to be about the size of the planet Jupiter, the density is believed to be about 355 times that of water—more than 45 times greater than steel.

Under Pressure

The force of pressure in the Sun's center is 450 billion times greater than on the surface of the Earth, and 100,000 times the pressure at the Earth's center. This pressure is equivalent to 3.3 billion tons per square inch, which is like saying that a postage stamp has to bear the weight of a mountain a mile high. If an average-sized automobile tire could bear this pressure—and if it could be supplied by an ordinary air hose at the gas station—the hose would have to be held at the valve for about 400,000 years.

Hot!

Temperatures in the interior of the Sun are approximately 16 million degrees Celsius (29 million degrees Fahrenheit).* A pinhead of this superhot material from the Sun's core would kill a person up to 100 miles away.

Higher and Hotter

Above the surface of the Sun (photosphere, "sphere of light") are its two basic layers of atmosphere, the chromosphere and the corona. The chromosphere, "sphere of color," is about 6,000 miles (9,654 kilometers) thick and is a brilliant red because of the hot hydrogen gas contained within it. Jetlike spears of gas, called spicules, constantly burn in this part of the solar atmosphere and give it the appearance of a "burning prairie," according to one astronomer. This layer has a temperature of about 20,000 degrees Celsius, some 14,000 degrees hotter than the Sun's surface. But with the next layer, the corona, the temperature jumps to an amazing 2 million degrees Celsius. In other words, the Sun's corona—its outermost layer of atmosphere, if the solar wind is not considered part of it—is a full 333 times hotter than the Sun's surface, which is like comparing liquid steel with warm bathtub water.

*Astronomers usually use Kelvin temperature, the scale of which starts at absolute zero, the lowest temperature possible, equivalent to −273 degrees Celsius. This book uses Celsius and Fahrenheit, but sometimes also uses Kelvin so that the reader can see the relationship between all three.

The Solar Wind

The solar wind blows a stream of high-energy particles off the Sun at velocities as high as 1.6 million miles (2.57 million kilometers) per hour, which could certainly qualify it as a solar "hurricane." These high velocities are reached when flares and prominences erupt on the Sun because of powerful magnetic forces. The more moderately paced streams of particles elsewhere on the Sun travel at about 670,000 miles (1,078,000 kilometers) per hour and might best be described as solar "breezes." The solar wind rushes and extends all the way to the planet Saturn—a distance of almost 900 million miles (1,448 million kilometers) from the Sun—and in one sense can be considered its outer atmosphere, which begins where the visible corona leaves off. Accepting this definition, the Earth itself is orbiting within the Sun's atmosphere. If the Earth's atmosphere were proportionately as large, it would extend far beyond the Moon.

The Solar Escape

The fastest escape velocity in the solar system is related to the largest mass—the Sun. To escape the surface gravity of the Sun, a speed of almost 1.4 million miles (2.25 million kilometers) per hour would have to be reached. This compares to the Earth's escape velocity of about 25,000 miles (40,000 kilometers) per hour and to the Moon's of 5,400 miles (8,700 kilometers) per hour. A speed of 93,600 miles (150,000 kilometers) per hour and a distance of at least 1 astronomical unit from the Sun are required to escape the gravitational dominance of the solar system. Since the Sun's speed around the Galaxy is estimated at 633,000 miles (1 million kilometers) per hour, the escape velocity of the Sun from the Galaxy would be

895,200 miles (1,440,000 kilometers) per hour. If we could travel as fast as the Sun's escape velocity, it would take only 11.8 seconds to travel from New York to San Francisco.

☆   ☆   ☆   ☆   ☆

THE SUN'S ENERGY

A Vast Melting

Enough heat from the Sun reaches the surface of the Earth in one year to melt a layer of ice 466 feet (142 meters) thick over the entire globe—almost 200 million square miles. If the Earth were made of ice, the Sun would melt it in 14,981 years.

Plenty of Energy

The Sun emits more energy in one second than mankind has consumed in the whole of its history. The Sun's power in watts is 380,000 billion billion kilowatts, which could run 176 billion billion frost-free refrigerators forever. One second of the Sun's energy is 13 million times the annual mean electricity consumption of the United States.

A Little Is a Lot

Even though the Earth receives only 2 parts per billion of the Sun's total energy output, this amount is nevertheless 10,000 times greater than the total energy presently consumed by the human race. If a George Washington quarter dollar represented the Earth, then a circular field with a mile-wide diameter would be the whole area that is reached by the Sun's radiation.

Energy Gold

The sunlight falling on 1 square yard of land in Arizona for a year, if converted into electrical energy, was worth $49.80 in 1960. The same amount of energy was worth about $83.00 in 1980. If all the sunlight falling on a 1-acre home lot in 1980 were converted into electricity, it would be worth $402,550!

The Hibernating Light

The light glow from a chunk of burning coal is really sunlight. Leaves on an ancient tree millions of years ago caught, transformed, and imprisoned some solar energy, and these leaves, given time and pressure, became coal. It is the combustion of the coal that releases the hibernating sunlight.

Whence Cometh the Energy?

Almost all the energy we use every day is converted solar energy. In fact, 99.98 percent of all energy passing through the atmosphere originates in the Sun's core. The other small fraction of 1 percent comes from starlight, cosmic rays, tidal energy, and geothermal energy.

Even though electrical energy may be produced by water-driven turbines, it is the Sun that raised the water from the earth to the height from which it rains or snows. Wood and coal is just stored-up solar energy—the coal taking millions of years to form. Wind and ocean-current power originate from the Sun's radiation, and, of course, so does the daily food we eat, which enables us to physically move about and, we hope, to reason and solve problems and survive. The Sun's energy comes to the Earth as a great gift that we are just beginning to accept and utilize with intelligence.

Direct to the Sun

A minute fraction of the vast power of the Sun produces responses in the human body—the squint reflex that protects your eyes when you walk from shade into bright sunshine or the definite heating up you feel when going from the coolness of a house interior to the bright midday sunlight of a sandy beach. That everyone's body has felt the power of the Sun makes it even more difficult to accept the fact that not much more than 2 parts per billion of its energy actually falls to Earth. Each square yard of the Sun's surface produces the equivalent energy of 70,000 horsepower (52,199 kilowatts). The amount of energy that falls on 1 square yard of earth on a sunny day would be approximately 1 horsepower (750 watts). In the sunny Southwest, a modest-sized garden of 49 square yards would receive about 222 kilowatt hours of energy each day, worth over $4,000 a year. This is a very modest example of how vast amounts of solar energy rain down on earth each day. The challenge in the next few decades will be to develop technologies that will economically collect, transmit, store, and convert solar energy into power for domestic and industrial use. If we go directly to the Sun, the ultimate source of all energy, we can eventually dispense with coal and oil and the services of their expensive middleman—250 million to 1 billion years of time.

The 500-Light-Bulb Postage Stamp

A piece of solar surface the size of a postage stamp constantly emits more energy than 500 sixty-watt bulbs and could light up all the rooms in 48 average-sized American homes.

God Bless Hydrogen

If the Sun had been composed of just oxygen and carbon, it would have burned for only 1,500 years and there would be neither people nor words for them to read in the present-day Universe.

The Sun's Diet

Deep inside the Sun, in its dense and hot core, nuclear reactions are converting hydrogen into helium, and energy is released in the process. Each second, in fact, the Sun loses 4.5 million tons of its mass by converting it into energy. Every 42 million years the Sun loses the equivalent mass of the Earth. If there were enough hydrogen, the Sun would use itself up and disappear in 14 trillion years. The Sun, however, is calculated to run out of hydrogen in about 3.5 billion years—and then start on the helium.

The Ultimate Efficiency

Just 1 pound of hydrogen that is converted into helium by nuclear processes in the Sun's core releases the same amount of energy as would 20,000 pounds (ten tons) of coal.

☆　　☆　　☆　　☆　　☆

THE SUN IN SPACE

The Sun from Afar

If we represented the Sun by a large 5-inch-plus (14-centimeter) orange, the Earth would be a sesame seed about 49 feet (15 meters) away and Pluto, a grain of millet, much smaller than the Earth, would be over 3,400 feet (600 meters) away. The nearest star to the Sun, Alpha Centauri, would be almost 2,500 miles (4,000 kilometers) away, according to this scale. The real distance happens to be somewhat over 25 trillion miles (4.3 light-years).

Imagine that we are on a planet that orbits Alpha Centauri. We look for the Sun in the night sky, and we see it between the familiar "W" of the constellation Cassiopeia and the stars of the constellation Perseus, not too far from the famous "Double Star Cluster" associated with Perseus. There it is, a yellowish star, almost as bright as Procyon or Rigel, one of the brightest stars in this planet's nighttime sky. But the Earth is hidden, never to be seen, not even through the most powerful telescope. The noise of our activities can be heard, however, through sensitive instruments.

The Sun's Odd Neighbors

While the Sun is often referred to as a "typical" or "average" star, after comparing a few of its features with neighboring stars within the very local neighborhood of 500 light-years, it is perhaps difficult to understand why. Hadar (Beta Centauri), 490 light-years away, is about 12,000 times brighter than the Sun. Acrux (Alpha Crucis), at a distance of 370 light-years, is a hot bluish star, its surface about 17,000 degrees Celsius (30,600 degrees Fahrenheit) hotter than the Sun's 6000-degree-Celsius (10,800-degree-Fahrenheit) temperature. Aldebaran (Alpha Tauri), the bull's eye of Taurus, is just a rocket thrust away at 68 light-years and is a middle-sized giant star, reddish-orange in color, that has a diameter about 800 times larger than the Sun and 87,334 times that of the Earth. Capella, the sixth brightest star in the sky and golden yellow in color, is really a multiple star system of four stars, all of which are about 45 light-years away. Sirius (Alpha Canis Majoris), the nearest naked-eye star visible from the United States, is a mere 8.6 light-years away from the Sun and has a famous, less conspicuous companion, Sirius B, a very small white-dwarf star, with a diameter about twice that of the Earth and a density about 90,000 times that of the Sun—where a cupful of matter would weigh over 5 tons. In the light of such company, how can we ever consider the Sun "average" again?

☆　☆　☆　☆　☆

SOLAR ERUPTIONS

Solar Prominences

Prominences are luminous fountains of bright hydrogen and helium that are expelled from the Sun and often rise to tremendous distances from the surface. They surge spaceward at high speeds, sometimes accelerating to 2.5 million miles (4 million kilometers) per hour. Just as often they shoot downward toward the Sun's surface or remain in one place in the corona, like luminous clouds. Their formation and motions are related to sunspots and powerful magnetic forces, and because of the various interplays of these forces, they assume many shapes—arches, columns, loops—and many sizes. Prominences may last for a few hours or as long as six or eight months. The average prominence would be 20,000 to 60,000 miles (32,000 to 96,000 kilometers) high, more than 100,000 miles (161,000 kilometers) in length, but only about 3,000 miles (5,000 kilometers) thick. The average volume of a prominence is 90 times that of the Earth. Like comet tails and rainbows, there's a lot to see but not much to hold on to. They are among the most beautiful and awe-inspiring sights in the solar system.

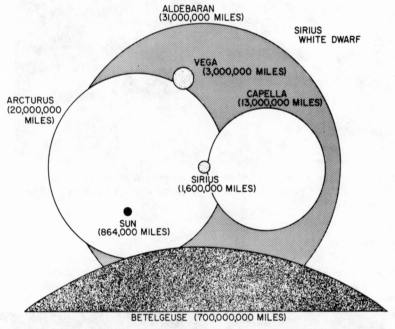

The size of our yellow dwarf Sun compared to some of its local neighbors within 520 light years. Courtesy Science Graphics, copyright © 1982.

A quiet prominence of 1971, where gas is suspended above the surface by magnetic fields. Courtesy Big Bear Solar Observatory, California Institute of Technology.

The great prominence of June 1946, which rose more than 1 million miles (1.6 million kilometers) above the Sun's surface—four times the Earth-Moon distance. Courtesy Harvard College Observatory.

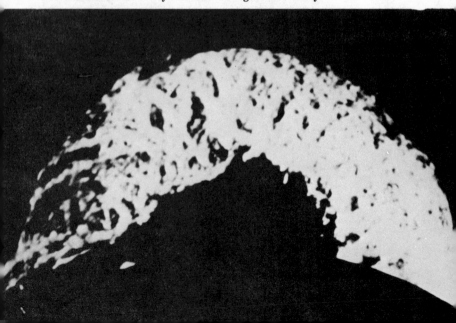

The Largest Prominence

The greatest solar explosion in recorded history occurred on June 4, 1946. This spectacular arch prominence reached a height of 250,000 miles (402,000 kilometers)—more than the distance from the Earth to the Moon—just a half hour after it was first observed. It continued to rise at a velocity of 400,000 miles (650,000 kilometers) per hour and soon went beyond the range of the camera. It probably rose more than 1 million miles (1.6 million kilometers) above the Sun's surface before dissipating. The entire eruption took only two hours, even though the prominence had been observed in its quiescent phase for several months before. One million miles may seem like an insignificant distance when compared to hundreds, even millions, of light-years, and so it does well to keep our perspective. Think of a cosmic necklace, made up of 126 Earths strung together; it would add up to 1 million miles, the height of this historic prominence.

A Solar Flash: Brighter than Bright

Solar flares, brilliant flashes of light in the solar atmosphere, are heard more than they are seen. While not as spectacular as the large prominences, their powerful eruptions expel gas and subatomic particles from the Sun across the vast distances of space at speeds upward of 1 million miles (1.6 million kilometers) per hour. When these high-energy particles interact with the Earth's magnetic field, they create aurorae (which few people see) and interrupt shortwave communication around the world—making themselves heard with static more than they are seen. Flares are associated with active sunspot groups, and their durations vary from four minutes to seven hours. One sunspot alone produced 80 flares as it rotated across the Sun's face. Flares are 10 times brighter than the Sun's normal surface, and they have been known to be as large as 1 billion square miles. The energies released from flares are tremendous: They are as powerful as a billion or more of the largest hydrogen bombs, and just one could scorch the Earth, making it a lifeless cinder.

Flaring Up

Solar flares and their high-energy protons are dangerous to astronauts' health, and it is hoped that these solar storms can be predicted in the future. Flares are likely to occur every eleven years when the Sun is active, but the most dangerous are those extraordinary and powerful flares that burst forth every few decades or so. Such violent eruptions send out deadly waves of powerful radiation. Just such an intense flare occurred in February 1956; it saturated the Earth with radiation and temporarily blacked out radio communications. If one of the Apollo crews had been on its way to or returning from the Moon during that time, all three astronauts would have been killed.

☆　　☆　　☆　　☆　　☆

SUNSPOTS

Sunspots in History

Chinese astronomers recorded seeing sunspots as early as 200 B.C., so these "dark" solar features have been observed for more than 2,000 years. A Chinese observer in A.D. 352 noted the following: "The Sun was a dazzling red, like fire. Within it there was a three-legged crow. Its shape was seen sharp and clear. After five days it ceased." About five to ten sightings per century have been found in ancient records. They were probably seen by the naked eye when the brilliance of the Sun was dulled during rising or setting, or when it was reddened by fog, haze, or smoke. Sunspots were not associated with the Sun itself by the early astronomers but were usually thought to be atmospheric features above the Earth or planets passing between the Earth and the Sun. Galileo studied them and recognized that the spots were part of the Sun. Since Galileo's time, scientists have used them to study the Sun's rotation by watching their movement across its face from day to day.

Formation of Sunspots

Sunspots—those cooler and darker blotches on the Sun's surface that follow definite cycles—are caused by the unique rotation of the Sun. Unlike the Earth, which rotates as a single solid unity, portions of the Sun, a great sphere of gas, rotate at different rates; this is called *differential rotation*. While the rotational period is 24.7 days at the equator, it is about 34 days near the poles. These different rotation rates for latitudes of the Sun cause the magnetic field to bend, stretch, and twist until it finally snaps and breaks through the solar surface. It is as if the Hudson Bay area rotated around the Earth slower than the Gulf of Mexico region, thus twisting and stretching the Great Lakes region and causing volcanic eruptions from deep within the Earth's interior. These magnetic ruptures—and their powerful dominant forces that completely inhibit the normal boiling of hot gases—are the sunspots that have held the keen interest of astronomers for centuries. The Sun's intense magnetic forces and their long-term configurations are therefore responsible for the various solar cycles, including the 11- and 22-year cycles that influence our daily life on Earth.

Sunspot Light

Sunspots are only dark in comparison to the brilliant sur-rounding solar surface. If an electric arc could be overlaid on the area of a sunspot, the electric arc would appear like a black dot in the sunspot. If the entire surface of the Sun could be blocked off in some way except for a single large group of sun-spots, the Earth would not be plunged into blackness. In fact, an orange-red glow would shine as bright as 100 full moons, and the naked eye would squint to protect itself from the daz-zling radiation.

The Cool Sunspots

Sunspots are cooler, and therefore darker, than the rest of the Sun's surface (photosphere). Their intense magnetic fields, which dominate the motions of matter in their vicinity, also inhibit the flow of heat into their area. The darkest and coolest part of a sunspot is about 4,000 degrees Celsius (7,000 degrees Fahrenheit)—still hotter than an acetylene welding flame—as compared to the surrounding blinding photosphere of nearly 6,000 degrees Celsius (10,000 degrees Fahrenheit)—about the temperature of an iron welding arc and 20 times the broiling temperature on a kitchen range.

The Largest Sunspot

The sunspot cycle reaches a "maximum" every 11.2 years, on the average, when the Sun is considered to be in its most active phase. The largest sunspot ever recorded appeared during such an active maximum in April 1947. It covered more than 1 percent of the area of the solar disk and was esti-

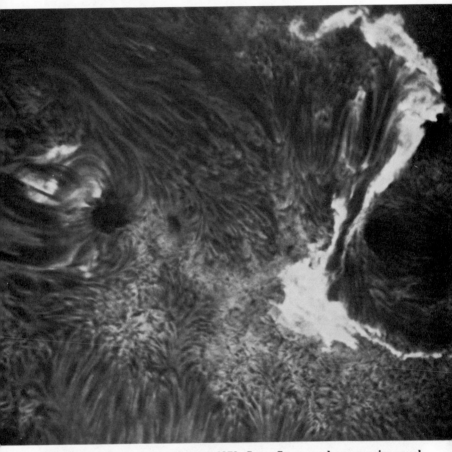

A flare surge near small sunspots, 1979. Some flares produce energies equal to 1 billion hydrogen bombs, and they cast off material at velocities up to 1 million miles (1.6 million kilometers) per hour. Copyright © 1979 Sacramento Peak Observatory.

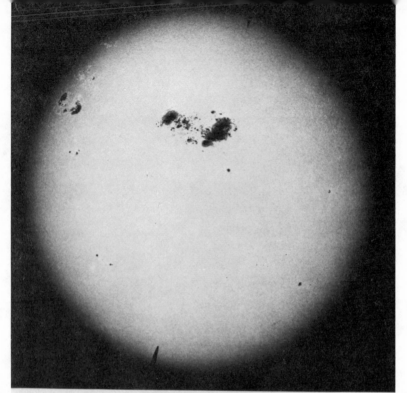

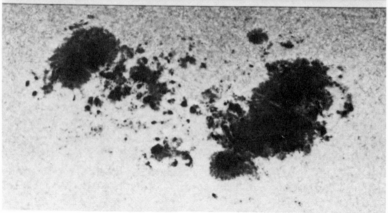

The largest sunspot group in recorded history occurred in April, 1947. Its area was 10 thousand times that of Alaska or large enough to contain about 100 Earths. Courtesy Big Bear Solar Observatory, California Institute of Technology.

mated to be about 6 billion square miles, over 10,000 times the area of Alaska—a gigantic turbulent solar region large enough to contain about 100 Earths that it could incinerate with just one super-powerful flare.

The Longest Sunspot

Most sunspots last only a few days, while a very few complete a solar revolution—about a month. The famous sunspot of 1840 holds the record as the longest lasting of these gigantic magnetic eruptions. It was visible for a full 18 months—220 times the duration of the worst recorded hurricane on the Earth.

☆   ☆   ☆   ☆   ☆
THE SUN'S CORE

Ultimate Evaporation

The Sun's core is so hot that a cube-shaped chunk of it, one mile on a side, has enough energy to melt the polar caps and then boil all the water of all the oceans, lakes, and rivers on the Earth.

The Amazing Neutrino

Neutrinos are subatomic particles produced by the nuclear fusion in the Sun's core. Along with gamma rays, they make up the energy released when hydrogen is converted into helium. Unlike gamma rays, however, which can take millions of years to reach the Sun's surface in the form of light and heat, neutrinos travel at the speed of light, reach the Sun's surface in about three seconds, and flash outward, speed undiminished, into interplanetary space. Neutrinos rarely interact with matter—they are practically unstoppable. A typical neutrino could travel through trillions of miles (perhaps several light-years) of lead shielding before being stopped. Without the lead, neutrinos would be able to travel forever.

The Sun's Secrets

The Sun keeps its big secrets locked away deep within its core, and until 1968 everything that science knew about star interiors, including the Sun's, was wholly theoretical. Scientists could learn from, and trust in, their studies of the Sun's surface, and all the electromagnetic radiations emitted from the surface—light, radio, x-ray, for example—were intensely studied, especially during times of high solar activity. But information from the surface told nothing of the Sun's core, since surface conditions resulted from nuclear reactions in the core a million or so years before. Then came knowledge of neutrinos—the only real-time connection science has with the heart of the Sun, since they pass through the Earth and us just over eight minutes after they are produced in the Sun's center. There finally was a way of checking the theory of our Sun's interior.

An astronomer, Raymond Davis, went about the task of catching and counting the elusive and all-but-impossible-to-stop neutrinos in the late 1960s. Deep underground (6,500 feet—1,981 meters, to be exact) near the South Dakota town of Lead, a vast tank was built and filled with tetrachloroethylene (more commonly known as cleaning fluid), a substance rich in chlorine atoms, some of which are chlorine-37. The nucleus of chlorine-37 will occasionally interact with a passing neutrino, producing a radioactive form of argon gas that can be detected. The extreme depth of the tank underground assured that the chlorine nuclei were not polluted by cosmic rays and other radiation.

Results from the big-tank neutrino catcher over a ten-year period showed a real discrepancy between the actual number of neutrinos that the Sun's core produces and the theoretical number that computer stellar models predict. This may mean that the present lower core temperature will cause a lower surface temperature—and a cooler Earth—in the future. If this is so, the big underground tank of cleaning fluid will have predicted a major Ice Age—about one million years in advance.

Through the Earth Darkly

A portion of the Sun's radiation continues to shine on us at night, when the Sun is on the other side of the Earth. Neutrinos, produced from nuclear reactions deep in the Sun's core, can travel through anything without being stopped, including the Earth. So, besides shining down on us during the day, neutrino radiation from the Sun also shines *up* on us at night.

Counting Your Neutrinos

The solar neutrinos that stream forth from the Sun amount to one fiftieth of the Sun's energy. Every second 100 million million of them shoot through your body—without one of your cells ever knowing it.

The Old Sunlight

Sunlight, traveling at the speed of light, takes just over eight minutes to travel from the Sun to the Earth, but the original radiation that created it is over 20,000 years old—older than recorded history. How can this be? Gamma radiation originating in the Sun's core is detoured millions of times as it collides with and bounces off densely packed solar particles on its tortuous journey to the Sun's surface.

Death Star

If the immense gamma ray energy (the shortest wave of all the electromagnetic radiations) generated in the Sun's core reached the surface (photosphere) unchanged, the Sun would be a dark Sun, radiating an all-encompassing death ray throughout the solar system and making life impossible. As it is, most of the gamma rays are dissipated in their long journey to the surface and become visible, infrared, and ultraviolet radiations that will, in turn, light, heat, and burn us.

ECLIPSES

Totally Off

The ancient Chinese chronicle called the *Shu Ching* contains a record of a total solar eclipse more than 4,000 years ago. The eclipse occurred on October 22, in 2137 B.C., and the Chinese populace was terribly surprised and frightened, because the royal court astronomers, Ho and Hsi, had failed to predict the eclipse accurately. With a precise prediction by their astronomers, the Chinese population could have prepared themselves psychologically with their traditional ceremonial displays of arrow-shooting and drum-beating, intended to chase away the dragon that swallowed the Sun. But this did

Total eclipses are rare and have been held in awe by many cultures. Records of them go back for more than 4 thousand years. This one occurred in 1918 over the state of Washington. Lick Observatory photo.

The eclipse that stopped a war. This resulted from Thales, the Greek philosopher, making his famous prediction. Courtesy Yerkes Observatory.

not happen in 2137 B.C. The chronicle relates that Ho and Hsi got drunk before the eclipse, and well they might, since their very lives depended on their accurate prediction: "Being before the time, the astronomers are to be killed without respite; and being behind the time, they are to be slain without reprieve." Ho and Hsi's worst fears came true. Their eclipse prediction was off, and so were their heads.

The Powers of Prediction

The Greek philosopher Thales, born in Miletus in about 640 B.C., correctly predicted the solar eclipse of 585 B.C. Much of Thales's fame in the Greek world was a result of this prediction, and he was honored as the chief of the Seven Wise Men of Greece. The total eclipse took place during a battle between the Lydians and Medes; the armies considered it a horrible omen and stopped fighting. A truce was signed the same evening, and eventually marriages between the warring parties brought a lasting peace.

Columbus was on the island of Jamaica for several weeks in 1504. He and his crew needed food because their store was low. The natives, however, were unfriendly. Soon after the Jamaicans refused to feed Columbus and his crew, Columbus noted that his almanac forecast an eclipse of the Moon. He warned the Jamaicans that his God would punish them by turning the Moon into blood. The natives scoffed at first, but when the eclipse occurred, they were terrified and promised him all the food he needed if he would just bring back the Moon. A deal was struck!

Today, eclipse prediction does not exert the same power over human behavior, but it is much more precise and long-term. For instance, there will be the rare seven eclipses in a single calendar year in 2160—two of the Moon and five of the Sun—about six generations from now. Pass it on, please.

Eclipse Behavior

During a total eclipse of the Sun, birds cease singing, animals head for their shelters or burrows, and insects fall asleep. Some people, especially those addicts who travel great distances to view a total eclipse, experience a temporary hysteria—and scream their heads off.

The Death of Totality

Total eclipses of the Sun are dying a slow death and will only be found in archaeological records a few billion years or so from now. The Moon is gradually spiraling farther away from the Earth because of tidal influences, and so the Moon's disk will never completely cover the Sun's disk in the far future (the requirement for a total eclipse). Future eclipse watchers will always see a ring of sunlight visible around the Moon, just as there is today when the Moon is at its farthest point from Earth and an annular eclipse occurs.

Lucky Relationship

The Sun and the Moon are apparently the same size in the sky, a fact that is hard to see for oneself, since the Sun should never be viewed directly. This fact is pure luck, an amazing coincidence, and makes possible one of the most spectacular and awe-inspiring events in the natural world—the total solar eclipse, when the Moon's disk almost perfectly covers the central body of the Sun. Two numbers are important in understanding this lucky relationship of size and distance: 400 and 108. The Sun has a diameter about 400 times that of the Moon; at the same time, the Sun is about 400 times farther away from the Earth than the Moon. Both the Sun and the Moon are about

108 times their own diameters away from the Earth. A round object of any size will have the same apparent diameter as the Sun or full Moon if it is placed a distance of 108 times its own diameter away from the viewer. Take a penny, for instance, and place it on a light-colored backdrop 81 inches (6¾ feet) away—the dot of the penny is the apparent size of the Sun or the Moon. This is hard to believe, especially when you stand squinting in the power of the Sun's glare.

Chase Scene

The Moon's shadow, a dark disk 167 miles (269 kilometers) wide, travels across the Earth at 1,040 miles (1,673 kilometers) per hour during a total solar eclipse. Duration of totality varies, but the longest possible time to observe the eclipsed body of the Sun from one location is 7 minutes 40 seconds. From a supersonic Concorde airliner traveling at the same rate of speed as the shadow, however, an observer could see totality for about 3.5 hours—a celestial chase scene in the name of science.

The Sun's Jewels

A total eclipse of the Sun provides a rare and fleeting display of the Sun's jewels, so the event has been called "the crown jewel of astronomical phenomena." First comes the diamond ring image, just before the Moon completely covers the Sun's disk, when only a single bright flash of sunlight remains, with delicate curves of sunlight emanating from both sides—looking just like a diamond in its setting. Next comes Baily's beads, the last sparkles of sunlight as the black edge of the Moon nears totality. This curved necklace of tiny bright points of light is created by the Moon's mountains and valleys breaking up the last sliver of sunlight. These are the Sun's precious jewels, seen only on these rare celestial occasions.

☆　　☆　　☆　　☆　　☆

THE SOLAR CONNECTION

A Shrinking Sun

The Sun's diameter—its horizontal measurement—is shrinking at about 5 feet (1.5 meters) per hour, and its vertical measurement is shrinking at about half that rate. Scientists John A. Eddy and Aram A. Boornazian believe that the Sun has been shrinking at this rate for the past 100 years, perhaps for as long as 400 years. It is theorized that this shrinking is a temporary contraction phase of a long-term oscillation pattern that might be providing the Sun, through gravitational contraction, a portion of its energy not produced by nuclear fusion. If the Sun has been shrinking at this rate for 100 years, it is almost 1,000 miles smaller in width than it was in the last century. If, as suspected, it has been contracting for 400 years, its outer layers have shrunk about 3,300 miles (5,310 kilometers)—more than the width of the United States.

The Pulsing Sun

The surface of the Sun pulses once every two hours and forty minutes (almost exactly one ninth of a day), moving in and out at a velocity of about 6 feet (2 meters) per second for a total change in the Sun's diameter of about 6 miles (10 kilometers). This recent discovery has shaken solar astronomy—which has had more than its fair share of upsets in recent years. These vibrations originate deep inside the Sun and appear to be a fast way for large amounts of energy to travel from the interior to the surface. The discovery was a surprise to solar astronomers, and their present theories will have to be reworked. Fingers are now crossed that the Sun has a regular, healthy pulse and will never skip a beat.

The Solar Weathermen

The space age has created a revolution in the study and knowledge of the Sun. Solar probes and above-the-atmosphere observatories (such as in Skylab) have allowed us to see and measure the Sun as never before. Every year more evidence accumulates for the Sun's direct influence over long- and short-term weather patterns on Earth. For example, there is a definite pattern of better weather during maximum sunspot activity and slower solar rotation; even the quality of wine vintages is superior during these periods. The relationship of bad weather to minimum sunspot activity and faster solar rotation is dramatically proven by the "Little Ice Age," which occurred during the late seventeenth century. From 1645 to 1715 there was an absence of solar sunspot activity. The Earth's temperature dropped during this seventy-year period by 0.6 degree Celsius (1.1 degree Fahrenheit), enough to cause the "Little Ice Age," during which European winters were so cold that the Thames River froze over ten times. It therefore seems likely that solar scientists will become important weathermen in the future, predicting severe cold spells and other long-term weather patterns by keeping a close watch on the Sun's rotation rate and the number of sunspots over a substantial period of time.

The Solar Holes

Recent solar research has found openings in the solar atmosphere. These openings, called *coronal holes,* are areas in the Sun's upper atmosphere, the corona, where the temperature and the density are lower than average, thus causing the magnetic force fields to open outward. The coronal holes allow the escape of solar material and are a source for high-velocity solar wind, which sweeps around the Earth, causing changes in our magnetic field. The frequency of these outpourings is related to the Sun's rotation, since the coronal holes rotate with it. Eventually scientists will be able to accurately predict when magnetic "storms" will arrive on Earth—extremely important information, considering humankind's increasing dependency on satellite communication for commerce and military purposes.

Taking the Sun's Temperature

Scientists began monitoring the Sun's temperature in 1975 and continue to take about six measurements per month. In January 1977 a drop in the Sun's temperature was noticed for the first time. This drop may be cyclical and related to sunspot activity; in fact, it changed when the sunspot activity passed its minimum and began to increase, and it appears to "track" well so far. The drop in temperature represents one tenth of one percent of the Sun's constant of 5427 degrees Celsius—9800 degrees Fahrenheit—about 0.5 percent decrease in the solar constant or energy output. Some climatologists have said a 2 percent drop in temperature over hundreds of years could trigger a new Ice Age, so the measured drop could be important if it is long-term. Measurements over a long period of time are needed before there is any definite proof that such small temperature changes influence the Earth's climate—and what we have to eat.

☆   ☆   ☆   ☆   ☆
FUTURE OF THE SUN

The Sun's Death

The realm of death extends even to the stars; our Sun will die. Astronomers, by following the rules of astrophysics, have created the probable death-throe images for all of us to see and contemplate some 6 billion years before the events take place. Such knowledge in the human mind might be considered the ultimate motivating force for humankind's continuing reach toward the stars.

The Sun is now middle-aged and healthy. Its earth-bound physicians pronounce a good average life for it as long as it has hydrogen to burn and convert into helium and energy. But in another 5 or 6 billion years there will be the ultimate energy crisis: The Sun's core will run out of hydrogen, contract, and begin to burn helium, converting it into carbon and oxygen; the Sun's atmosphere, at the same time, will expand at a tremendous rate to 100 times its present size, and the Sun will become what is known as a red giant. This old-age phase presents the most dramatic and frightening image of the Sun's future. This giant red Sun will swell out to, perhaps beyond, the Earth's orbit, melting the continents, boiling away the oceans, killing any life left on the surface, vaporizing our entire planet. The red-giant phase will last for a few million years, and then gravitational forces will begin the final collapse, as the Sun ejects great waves of matter, shaking with cosmic death spasms. The Sun will finally contract to about the size of the Earth and become a white-dwarf star of intense density, where a handful of matter can weigh 16 tons. Slowly, very slowly, the white dwarf will radiate the last of its heat into space, cool off, and die, until perhaps another supernova blows its remnants into a new batch of star stuff from which a new star may be born.

All or Nothing

The future survival and quality of human life depends on future energy resources. The ultimate answer—seemingly far-fetched now, perhaps, but not in a few thousand years—will be to utilize the *total* energy output of the Sun (the Earth receives only 2 parts per billion of this total output). While direct use of all the solar energy that falls on the Earth each second (almost 100,000 times more than today's production of all forms of energy) seems like energy riches untold, it is estimated that all this terrestrial solar energy will fall far short of the energy needs for A.D. 4500—short by a factor of 100,000.

The famous American physicist Freeman J. Dyson has done some exciting theoretical work on an ultimate solution to this problem. Dyson proposes that humankind will be capable in 2,500 to 3,000 years of building an immense shell that will surround the Sun, a structure that has become known as the Dyson sphere. On the inner surface of this sphere, which would be built by using the entire mass of the planet Jupiter, would be an artificial biosphere where future civilization can thrive. This magnificent star sphere would allow humankind to use the total energy output of the Sun; every photon of it can be absorbed and used. The Dyson sphere would have an inside surface area 1 billion times greater than the Earth's surface area, so crowding would not be an immediate problem.

Moving the Sun

Far-future speculators (notably Adrian Berry in *The Iron Sun*) describe the possibility of creating a black hole about 1 light-year from the Earth by magnetically bulldozing vast amounts of interstellar dust into a mass 10 times the weight of the Sun (3 million times the weight of the Earth). This black hole would form one fourth of an interstellar bridge to make possible instantaneous starship travel across immense distances of space and time. Along with a neighboring white hole (the arrival port for the return trip) and a distant white hole and black hole (arrival and departure ports, respectively), there would be established a round-trip interstellar travel route. In so doing, however, the Sun would lose its status as a single-star system rotating around the Galaxy and become a three-star system, since the black and white holes about 1 light-year's distance would change the local gravitational fields. This, in turn, would cause the Sun to move differently in relation to the other stars, and all the star atlases and planetarium projectors would be slightly off. In effect, the Sun would have been moved.

No Escape?

The Sun would become a black hole if its diameter of 865,000 miles (1,392,000 kilometers) were somehow magically compressed to just under 4 miles (6 kilometers). But compression would not stop there. Further collapse would reduce it to an infinitesimally small size. Would intense gravitational pull wrench the Earth from its orbit, swallow it up, and annihilate all its matter? Would there be no escape for the Earth, since not even light can escape from these immense forces, and light travels at 670 million miles per hour? Actually, nothing at all would happen to the Earth's orbit; it would just become an orbit around a black hole rather than around a yellow dwarf star—well out of the black hole's gravity reach.

Cooking with Sunlight

The kitchens of future space habitats may not have gas ranges or electric ranges but, instead, Sun ranges, on which the Sun's radiation will be directly used for cooking. The solar heat will never be more than a few feet away if the residence is attached to the habitat's interior wall. The Sun's heat can thus be brought up from a collecting mirror, through the dwelling's floor, and directly onto the metal cooking surface. A shutter would be opened or closed to turn on or off the Sun range. The cook's pride might be in serving up Sun-roasted chicken or Sun-baked apple pie.

Sailing the Solar Wind

Sailing spaceships, their great sails propelled by the Sun's radiation, may appear in future night skies like brilliant stars. The Sun's enormous energy that radiates throughout the entire solar system exerts a force on everything it shines on and reflects from. This is true for all light; even a flashlight exerts a force against its casing when turned on, but it amounts to a mere one ten-trillionth of a pound. Given a large-sized square area and some time, this same light force can create powerful and practical accelerations, able to ferry payloads around the solar system. A mylar solar sail of 10,000 square feet (929 square meters) can reach a velocity of 2,500 miles (4,022 kilometers) per hour. In about six weeks it can reach a velocity comparable to the Apollo spaceships—25,000 miles (40,000 kilometers) per hour. The greatest advantage of these sailing spaceships is that their fuel—the Sun's radiation—is free, plentiful, and constant once they are boosted into Earth orbit. The trick will be the unfolding of the sail—perhaps 10,000 to 20,000 feet (3,050 to 6,100 meters) of it—from a small canister not much larger than a trash barrel.

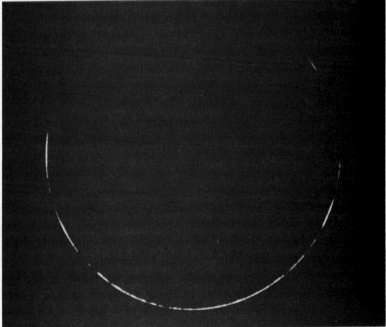

The Sun's jewels: (Top) diamond ring effect of 1966 eclipse and (Bottom) Bailey's beads of the 1952 eclipse, when the lunar mountains break up the delicate necklace of light. Courtesy NASA and Yerkes Observatory.

One of the first close-ups taken of Mercury by Mariner 10 in 1974.
Photographic resolution of the planet increased 5-thousand-fold in just one
year. Courtesy NASA.

THE OTHER PLANETS: THE MAGNIFICENT EIGHT

☆　☆　☆　☆　☆

MERCURY

The New, Revised Mercury

The planet Mercury had many claims to fame for the first seventy years of the twentieth century. It was the innermost planet of the solar system and therefore the fastest; it was the smallest; since it was closest to the Sun, it was also the hottest planet; and, like the Moon, one side constantly faced the Earth, locked there by the Sun's gravitational power.

Then discoveries in the 1960s and '70s dramatically changed some of these facts about Mercury, first through radar studies with the great Arecibo radio telescope in Puerto Rico and later, in 1974, with the data and photographs collected by Mariner 10, the first space probe with a dual-planet (Venus and Mercury) mission. All the textbooks had to be rewritten, and Mercury lost several of its claims to fame. It became the first of the new, revised planets.

Mercury did retain two important titles; it still was the closest planet to the Sun, and it still had the fastest orbital velocity. But in the late 1970s, Pluto, not Mercury, was found to be the smallest planet, with a diameter of about 1,550 miles (2,500 kilometers) versus Mercury's 3,030 miles (4,880 kilometers). Then the planetary probes proved that Venus had the hottest surface, because of its thick atmosphere and the greenhouse effect, taking another title away from Mercury. Another blow to the first edition of Mercury came when radar research in the 1960s, followed up by additional probe information, told astronomers that they had been wrong for almost a century—Mercury did indeed rotate very slowly, once every 58.65 days, exactly two thirds of its orbital year.

The Mariner 10 planetary probe provided over 10,000 close-up photographs of the planet during its three encounters, and so the new, revised Mercury was heavily illustrated. Before 1974, Earthbound telescopes provided just fuzzy images of the planet going through its phases, with a few indistinct markings. Photographic resolution of Mercury's surface features thus increased 5,000-fold in just one year.

Swift Mercury

Named after the speedy messenger of the gods, Mercury is the fastest of all the planets as it whips around the Sun once every 88 days, which also gives it the shortest year in the solar system. At its farthest distance (aphelion) from the Sun, Mercury moves along at over 87,000 miles (140,000 kilometers) per hour, but accelerates to a top speed of about 127,000 miles (205,000 kilometers) per hour when it is closest (perihelion) to the Sun. Mercury's increase of orbital velocity of 40,000 miles (65,000 kilometers) per hour is, however, done over a 44-day period—half its orbital year—speeding up 909 miles (1,477 kilometers) per day or 37.8 miles (61.5 kilometers) per hour, which is not really all that fast. It is about 1,000 times less than a sports car accelerating from 0 to 60 in six seconds.

Surfer's Delight

The surface of Mercury resembles that of the Moon so closely that even experts can be fooled by photos. But if Mercury were to replace our Moon in orbit around the Earth, there would be no mistaking the difference. It would be about 1.5 times as large in diameter and would shine with twice the light. The most dramatic change if we had Mercury orbiting Earth would be higher tides because of Mercury's greater mass—4 times higher. Coastal areas of the world would be awash, and daring surfers would ride their boards where once there were freeways and traffic jams.

Faint Echoes from Mercury

Reflected radar signals from Mercury in the 1960s determined night-side temperatures of the planet as well as Doppler shifts, which eventually led to the discovery that Mercury does, after all, rotate ever so slowly—about once every 59 days.

The great Arecibo radio telescope, wider than three football fields, with its dish in the natural contour of the Puerto Rican hills, beamed its powerful expanding signal toward Mercury and received the minute echo portion about 15 minutes later, after a round trip of over 180 million miles (290 million kilometers) at the speed of light. The feeble echo actually received amounted to only one million-trillionth of the tiny reflected portion of the original signal sent. In fact, the full amount of energy that fell on the Arecibo dish during the entire

Mercury radar research program would not even light a 100-watt bulb for ⅟₅₀₀,₀₀₀ of a second.

A Mercury Calendar

With its slow rotation and fast orbit, there are 1½ Mercury days in every Mercury year, 6.2 Mercury days in every Earth year, and just over 4 Mercury years in each Earth year. This means that men landed on the Moon less than 100 Mercury days ago, and that one U.S. presidential term lasts just 25 Mercury days—an example of how planetology can offer optimism to many people.

Mercury's Thin Helium

An extremely thin envelope of helium gas surrounds Mercury, but this in no way can be considered an atmosphere. Whatever atmosphere Mercury once had was lost billions of years ago because of the weak surface gravity, about one third that of the Earth's. The tenuous helium is believed to come from the solar wind and is trapped by Mercury's magnetic field. The amount of helium above Mercury is so sparse that all of it in a 4-mile (6.4-kilometer) diameter sphere would just be enough to fill a child's small balloon.

Sunrise, Sunstop, Sunset

The motion of the Sun in Mercury's sky is peculiar, to say the least, when the planet is at its closest distance to the Sun. The sight is visible nowhere else in the solar system.

From a location on Mercury's largest geographical feature, a huge plain called the Caloris Basin, observers would see the Sun (over 3 times larger than as seen from the Earth) rise in the east, move slowly across the sky, then slow down and come to a complete halt when almost due south. Next the Sun would move westward, and then finally set in the west.

The observers would have to be protected from the broiling temperatures of Mercury's daytime. They would also have to muster almost inhuman patience, since Mercury's strange sunrise, sunstop, and sunset would take 88 Earth days.

Extreme Degrees

Even the coolest sunny noontime of Mercury, when the planet is at its farthest point from the Sun, reaches a temperature of 285 degrees Celsius (545 degrees Fahrenheit), hotter than the highest temperatures of most kitchen ranges. When Mercury's orbit takes it closest to the Sun, noontime temperatures climb to a sizzling 430 degrees Celsius (806 degrees Fahrenheit), almost 4 times the boiling temperature of water and 7.5 times the hottest recorded temperature on Earth—57.7 degrees Celsius (136 degrees Fahrenheit), observed at Azizia, Libya, in Northern Africa, September 1922.

Nights on Mercury can reach −180 degrees Celsius (−292 degrees Fahrenheit), 7 times colder than the normal freezer compartment of a refrigerator. So Sun-hugging Mercury has the greatest temperature range of all the planets—over 600 degrees Celsius (1,100 degrees Fahrenheit). At its hottest, it could melt lead or tin, or turn steel red hot in just minutes; at its coldest, it could freeze a person solid in just minutes.

Where the Sun Never Shines

In Mercury's polar regions, there are small craters with such steep walls that their floors have been in chilly darkness ever since their formation. These so-called cold traps may be prime landing sites for any future Mercury space probes because they are preserves for the planet's primordial substances. Volatile substances such as carbon dioxide and water that escaped billions of years ago from Mercury's interior may remain trapped in the form of ice to this very day. If this rare and ancient ice is ever scooped up and sampled by an unmanned probe, it will be brought into a special compartment, allowed to melt, and then analyzed. This would probably be the first water on Mercury's surface in billions of years.

The Iron Planet

After the revelations of the planetary probes, Mercury was generally described by some astronomers as a planet whose surface looked like the Moon, pockmarked with craters, but whose interior was similar to the Earth's. In fact, Mercury has a proportionately larger iron-nickel core than does the Earth—it amounts to 80 percent of the entire planet's mass, with a diameter of at least 2,236 miles (3,600 kilometers). This makes Mercury's iron core slightly larger than our Moon. At recent world production rates for iron, it would take 650 billion years to mine all the iron in Mercury's core.

Weighing in Mercury

The surface of Mercury is equal to 15 percent of the Earth's surface, about the surface area of North America. Its mass is only 6 percent of Earth's, so it would take 16.6 Mercurys to equal 1 Earth. If Mercurys were piled up on one side of a cosmic scale to equal the Sun's mass, over 5.5 million Mercurys would be needed to balance the scale.

Counting Calories in Caloris

The Mariner 10 spacecraft photographed about half of Mercury's surface during its three approaches in 1974 and 1975, and the largest impact feature discovered was Caloris Basin, with a diameter of over 850 miles (1,400 kilometers)— about one fourth the entire diameter of the planet or a distance greater than from New York to Chicago.

A large asteroid over 60 miles (100 kilometers) in diameter blasted into the Mercurian surface at a speed as high as 318,000 miles (512,000 kilometers) per hour. Such tremendous velocity is possible because of Mercury's lack of atmosphere and its closeness to the Sun, where the more powerful gravitational forces cause in-falling objects to impact more violently than they would on Earth.

The colossal explosion shook the entire planet and may in fact have created the hilly and peculiar terrain on the opposite side of the planet. It is believed that the impact caused subterranean molten lava to gush up from the interior through thousands of cracks and fissures that were blasted open, some as wide as 5 miles (8 kilometers), equal to the width of the Grand Canyon in many places. The lava spread out and left the relatively smooth plains seen covering the gigantic basin.

The crater basin was named Caloris because of its location in the extremely hot equatorial zone. If the tremendous energy of the blast were measured in calories, assuming the high velocity, it would have produced enough calories to sustain the world's population, 4.5 billion people, for 360 million years at 2,000 calories a day. If these 1.2 trillion trillion calories were made into a gigantic banana split, the dessert would be 544 miles (875 kilometers) high, with an extra 300 miles (483 kilometers) for the whipped cream, and a cherry 80 miles (129 kilometers) wide.

The Cliffs of Mercury

Despite the apparent similarities between Mercury and our Moon, each has unique geological features. Mercury, for example, is covered with unusual massive cliffs that are over 300 miles (500 kilometers) long and as high as 2 miles (3.2 kilometers). These cliffs are geologically young because they cut across the ancient basins and craters and the smoother lava-formed maria. Called *lobate scarps* by planetary geologists, these cliffs are much steeper on one side than the other, and their crests are rounded. Scientists believe they represent faults that were thrust up during the compression of Mercury's huge iron core billions of years ago. Such crustal movements indicate that the entire planet is shrinking, and that it is smaller today than it was in the early solar system.

Santa Maria Rupes is an example of such a cliff range on Mercury and rises about 10,500 feet (3,200 meters) above the surrounding surface. This is 2,000 feet (610 meters) higher than the highest sea cliff on Earth, located on the east coast of Molokau, Hawaii. In Mercurian gravity, a rock thrown off this cliff would take 2.2 minutes to hit the ground at about 109 miles (175 kilometers) per hour, almost 3 times slower than on Earth because of the weaker gravity.

Mercury's Fate

Mercury will be the first planet to be destroyed in the solar system—assuming, that is, that humankind gets its survival act together. As the Sun swells and becomes a red giant about 5 billion years from now, innermost Mercury will evaporate as it is engulfed by the Sun's expanding outer layers. There will be no escape for swift Mercury—born of, and claimed by, the Sun.

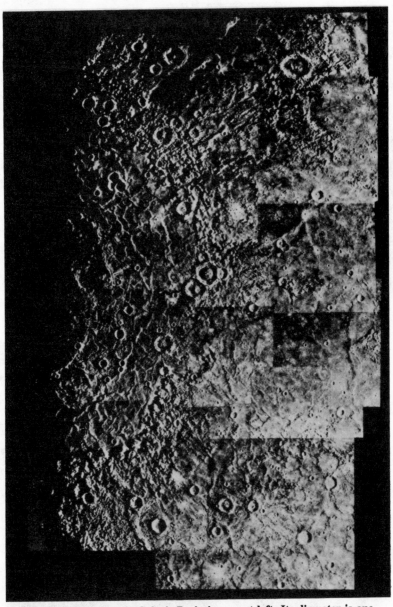

A portion of the gigantic Caloris Basin is seen at left. Its diameter is one-fourth the entire diameter of the planet—a distance greater than New York to Chicago. Courtesy NASA.

Venus is just a fuzzy crescent through telescopes on Earth, but Mariner 10 gave us the first view of the sulfuric acid cloud swirls in 1974. Courtesy C. F. Capen, Braeside Observatory and NASA.

VENUS

Bright Venus

Venus is brighter than any other planet or star (excepting the Sun, of course) in the sky. At certain positions it is 12 times brighter than Sirius, the brightest star, and can actually cast shadows on Earth. The Greek astronomer Simplicius referred to the shadows cast by Venus in the sixth century A.D. Under ideal conditions it can also be seen during daylight; Napoleon Bonaparte saw it shining at noon one day in Luxembourg.

Like the Moon and Mercury, Venus goes through phases as viewed from Earth. Surprisingly, when we see Venus at its brightest, it is not in its full phase. Rather, it is in a crescent phase at its closest point to Earth—about 25 million miles (40 million kilometers), and closer to Earth than any other planet. At its full phase, Venus is beyond the Sun, as far as 160 million miles (257 million kilometers) away, and this great distance makes its apparent diameter only one sixth of what it is at its close crescent phase. This is why we see the brightest Venus not when it is full but when it is a nearby crescent. If Venus did have its full phase when closest to the Earth, its extra light still would not compete with moonlight, which would be 175 times brighter.

The Nearest Planet

When closest to the Earth about every 20 months (584 days), Venus is some 25 million miles (40 million kilometers) away. Today, after the American and Russian planetary probes and all the revelations they provided about the environment of the shrouded planet, this proximity of Venus and Earth appears to be the only reason left to continue the ancient tradition of calling them the "sister planets."

Even at this close distance, Venus is still about 106 times farther away from us than the Moon. At the average speed of an Apollo spacecraft, a one-way manned mission to Venus would take 1.5 years. If you wanted to travel to Venus and obey terrestrial speed limits in order to conserve fuel, it would take 52.7 years just for the one-way trip at 55 miles (88 kilometers) per hour.

The Longest Day

Venus rotates on its axis once every 243 Earth days, a fact that was not established until 1961, when radar technology passed through the thick obscuring clouds and measured the motion of the planet's surface. Its period of revolution around the Sun was known for much longer, however—Venus orbits the Sun every 225 days. This means, based on the Earth system, a day on Venus is longer than a year—a unique relationship in our solar system. World War II thus ended 54 days ago, according to the Venus calendar.

Contrary Venus

On Venus, the Sun rises in the west and sets in the east, which is to say that it rotates east to west, unlike all the other planets of the solar system, unless oddly oriented Uranus is considered a second exception. This is called *retrograde motion*. The rotation of Venus is therefore unique in two ways— in its extremely slow rate, equal to about eight Earth months, and its genuine reversed direction, relative to the eight other planets.

How did oddball Venus come to rotate east to west? Some astronomers propose that Venus had a large moon early in its history (like Mercury, it has none now), which once revolved around it in retrograde motion. This satellite crashed into Venus and initially stopped, perhaps even reversed, the planet's direction of rotation. Another theory, perhaps more likely: Venus got its odd rotation during its formation in the early solar system, when two very large rotating planetoids collided off-center in such a way as to cancel out their rotation. If its early rotation was similar to the Earth's, then the energy required to reverse it would have been tremendous—equal to more than 8.3 minutes of the Sun's total output or the amount of electricity the United States could generate, at current rates, for the next 15 billion years.

Venus Unveiled

The first definite radar contact was made with the planet Venus in 1961, and Earth-based radar studies have continued up to the present time on small portions of the planet. But Earth-based radar depends on a close Earth–Venus distance, so that the radar echo is sufficiently strong to supply valid data. Also, since Venus's rotation rate is so slow (243 days), the same basic face, with just a change of a few degrees, is observed with each close approach. For this reason, Earth-based radar studies of Venus are severely restricted; it would take over 1,000 years to cover the planet's 360 degrees of longitude. After one decade, less than 1 percent of Venus's surface had been mapped by Earth radar. Still, it was the beginning of Venus's unveiling, a planet cloaked with thick, impenetrable clouds that had frustrated humankind's curiosity about it ever since its brilliant night-sky light captured our imagination.

Then the Pioneer-Venus spacecraft went into orbit around

the planet in December 1978 and in less than two years mapped 93 percent of the Venusian terrain, providing us with the first global lay-of-the-land view of Venus. It discovered great mountains, canyons, rolling hills, plateaus, and low-lands—all the large-scale planetary features. Only the north and south pole regions were not covered. The radar mapper also measured the terrain's general surface roughness on a scale down to 1 meter or so. More was mapped of the planet Venus in two years than had been mapped of planet Earth up to the year 1800.

The Venusian Winds

The atmosphere of Venus extends to an altitude of about 155 miles (250 kilometers) above the scorched surface, and the whole system blows westerly around the planet, circulating about once every four days. The winds at the highest cloud layer reach 220 miles (355 kilometers) per hour, a velocity close to Earth's jet streams and the record surface wind speed clocked at Mount Washington, New Hampshire. The middle cloud layer, the fastest on the planet, attains speeds of almost 450 miles (724 kilometers) per hour, which is substantially faster than the highest measured tornado speed on Earth, 280 miles (450 kilometers) per hour. Wind velocities drop off considerably at the bottom layer of clouds to about 100 miles (160 kilometers) per hour, and to a calm 2.24 miles (3.6 kilometers) near the surface of the planet.

The calm of the surface breeze is illusory, however, since it is a breeze of fire, about 470 degrees Celsius (878 degrees Fahrenheit), which can split rocks and destroy well-planned space probes in a matter of minutes. Add to the inferno the dense pressures that exist on Venus, and you have a surface breeze that could level all the wood-frame dwellings on Earth in just minutes.

Acid Weather on High

The three distinct cloud layers of Venus are all composed of sulfur, although the density and size of the particles vary. The largest sulfur particles are in the lowest layer, which is considered the main cloud deck, at an altitude of about 30 miles (49 kilometers). Here the particles are about the size of large raindrops. In the middle cloud layer there are acid droplets, smaller in size, as well as some liquid and solid sulfur particles. The highest cloud region, from about 38 to 43 miles (62 to 70 kilometers) above the surface, is composed entirely of sulfuric acid droplets, smaller still, about the size of fine dust. This sulfuric acid is more concentrated than in a car battery. Within these high-flying acid clouds, the visibility would be about 0.6 mile (1 kilometer), more like a light haze or smog than the conventional opaque clouds on Earth.

While the yellowish sulfur haze extends even below the main cloud deck, down to an altitude of about 20 miles (32 kilometers), the atmosphere then becomes clear and remains so all the way to the blistering surface.

The surface of Venus has no weather in the terrestrial sense—no large fluctuations of temperature, wind, pressure. The real weather of Venus begins in its cloud layers, about 20 miles (32 kilometers) high, which would be like the Earth's weather system beginning in our upper stratosphere, where actually no clouds, no weather conditions whatsoever, exist.

Where Has All the Water Gone?

Small amounts of water vapor were detected below the Venusian clouds by the Pioneer-Venus probes and were measured at between 0.1 and 0.5 percent of the massive atmosphere. This is a distant third place from dominant carbon dioxide (about 96 percent) and nitrogen (3.4 percent), but is nevertheless more than many scientists expected after examining data from the earlier Russian probes. Even though Venus's water vapor is less than 1 percent of its atmosphere, it is still a significant factor in trapping the heat and causing the runaway greenhouse effect, since water absorbs infrared radiation so efficiently.

Why is Venus so dry, especially above the cloud layers? Many scientists believe that primordial Venus had as much water as Earth. If so, where has all the water gone?

One explanation is that Venus was originally hot enough (owing to its being closer to the Sun than Earth is) for all its water to evaporate into the atmosphere. Because this hot and waterlogged atmosphere had no cold trap similar to the Earth's stratosphere, the water broke down into hydrogen and oxygen, and the light hydrogen atoms leaked away into space. The oxygen then may have reacted with the surface rocks to oxidize them. The water, in essence, was driven away by heat. The water left on Venus today would equal about 270 trillion tons, enough to cover the surface to an average depth of about 2 feet (0.6 meter)—a planet-wide pond.

An Extinguishing Atmosphere

The atmosphere of Venus, over 40 times more massive than the Earth's, is composed of about 96 percent carbon dioxide. This is equal to the mass of a large asteroid or a small moon some 300 miles (500 kilometers) in diameter. It is enough carbon dioxide to fill about 100 million trillion fire extinguishers.

The Bolts of Venus

American and Russian space probes to Venus have detected gigantic lightning storms near the surface of the planet.

The Soviet Venera 11 detected as many as 25 lightning impulses each second as it descended to the surface on December 21, 1978. Then, on Christmas Day, Venera 12 recorded 1,000 impulses in less than 4 miles (6.4 kilometers) of its descent path, between the altitudes of 6.8 and 3 miles (11 and 5 kilometers) above the Venusian surface. Early in 1979 the Pioneer-Venus spacecraft orbited as low as 88 miles (142 kilometers) above the night side of the planet, and signals were registered immediately—signals that most scientists, after studying their spectra, believe are generated by lightning.

On Earth, about 1,800 thunderstorms, along with hundreds of lightning strokes, take place at any one time. But just one small area of Venus would have hundreds, perhaps thousands, more bolts than the entire Earth. So the planet of extremes is crackling as well as sizzling.

The Smoldering Greenhouse

The inhuman surface temperatures on Venus, higher than on any other planet, probably vary no more than 20 to 30 degrees planet-wide. The temperature range is found in the altitude above the surface rather than any regional (longitude and latitude) differences.

At 470 degrees Celsius (878 degrees Fahrenheit), the temperatures of the surface would turn lead molten and still have 100 degrees left over to make steel red hot. Even at a height of 30 miles (49 kilometers), where the dense clouds begin, the temperature drops only to 70 degrees Celsuis (158 degrees Fahrenheit). At about 42 miles (68 kilometers) above Venus, however, the temperature finally drops to the familiar 0 degrees Celsius (32 degrees Fahrenheit).

When Earthbound radar studies in the 1960s indicated such high temperatures, many scientists were skeptical. The results, however, were verified by the increasingly sophisticated Venus space probes of the 1970s. The planet Venus, it turns out, is a greenhouse without a keeper, whose temperature control has broken down. Sunlight comes in and heats up the atmosphere and surface, but the heat is then trapped; very little can escape through the dense atmosphere of carbon dioxide which makes up about 96 percent of the planet's atmosphere. This process has been called the "runaway greenhouse effect." But Venus, almost as large as Earth, is a giant inferno, too hot to be a greenhouse for any living thing. Its surface heat would incinerate the Earth's billions of trees and other greenery immediately.

Lightfall on Venus

At an average distance of 67.2 million miles (108.2 million kilometers) from the Sun, Venus is about 26 million miles (42 million kilometers) closer than the Earth. Its distance varies the least of any planet because of its nearly circular orbit.

Despite Venus's nearness to the Sun, only slightly more than 2 percent of the original sunlight striking the planet is actually absorbed at the surface. This compares to about 50 percent for the Earth.

First, high above the surface, 70 percent of the sunlight is reflected back into space by the upper cloud layer. Another 15 percent is absorbed by the haze above an altitude of 37 miles (60 kilometers), and 12 percent more is absorbed in the main cloud deck. What emerges into the clear atmosphere at a height of about 20 miles (32 kilometers) is the remaining 3 percent, which then falls toward the dark and low-reflective (15 percent) surface.

While 1 hour of sunlight falling on about 11 square feet (1 square meter) of the Venusian surface could light a 100-watt bulb for a half hour, sunlight falling on the same area of Earth (specifically, in the southwestern United States) could light it for 8 to 10 hours. Surface heat, of course, is another matter. Venus has more heat energy per square mile (2.6 square kilometers) at its surface than Earth does in 2.57 square miles (6.66 square kilometers).

Under Pressure on Venus

No person could survive the hellish atmospheric pressure on the surface of Venus, 90 times that of the Earth's. While the Earth has a pressure of 14.7 pounds per square inch (6.45 square centimeters), sister Venus exerts an oppressive 1,320 pounds on this same small square area—a pressure equal to what a diver would experience at 264 feet (80 meters) below the ocean surface, or that a submarine hull would withstand at an ocean depth of 3,280 feet (1,000 meters).

If a suicidal astronaut (probably an all but impossible combination) landed on Venus and opened the hatch of his thick-walled ship—half rocket, half submarine—his lungs would burn up, as if he were inhaling the hot gases of a volcano, before he could smell his first whiff of sulfur.

Corrosive Rains

The sulfuric acid droplets in the high Venus cloud layers may sometimes fall on the planet's surface. As a sulfuric acid raindrop fell through the increasingly hot and dense atmosphere, the water would evaporate and the acid would become more concentrated. At times, depending on atmospheric chemistry, the acid could react with hydrogen fluoride and form fluorosulfuric acid, a strong mineral acid that can dissolve mercury, lead, tin, sulfur, and most rocks. Such a corrosive rain on Venus would be one of the most potent and destructive fluids in the solar system. It would burn away human flesh in a matter of minutes. Umbrellas would not help at all.

The Ups and Downs of Venus

The seething surface of Venus for the most part is gently rolling, with a few dramatic highs like North America. These common rolling plains comprise about 60 percent of the planet's newly radar-mapped surface and define what would be the equivalent of sea level of Earth, but which would better be called "plain level" on Venus—a surface close to the planet's mean radius of 3,758 miles (6,051 kilometers). This large Venusian area varies no more than 300 feet (91 meters) in height.

Large shallow basins and valleys make up the lowlands—those areas below "plain level"—and together they account for 16 percent of the mapped surface. The lowest point is 9,500 feet (2,900 meters) below plain level. Another 24 percent of Venus's surface is on average 3,280 feet (1,000 meters) above plain level, 8 percent of which can be considered true highlands, with a maximum at the summit of Maxwell Montes.

If we do not count the Earth's ocean depths, Venus has a greater range of elevations than our home planet. The largest difference in dry-land elevation on Earth is measured from the summit of Everest to the deepest point on the bed of the Dead Sea—31,628 feet (9,642 meters). On Venus, it is 44,900 feet (13,689 meters) from the summit of great Maxwell Montes to the lowest depression. The difference in elevation extremes between Venus and Earth is therefore over 13,000 feet (3,963 meters), which is about the average altitude for the Swiss

Maxwell Montes, one of the greatest mountains in the solar system, is over a mile higher than Mount Everest. Lightning bolts probably crack around its summit. Artist's concept courtesy of NASA.

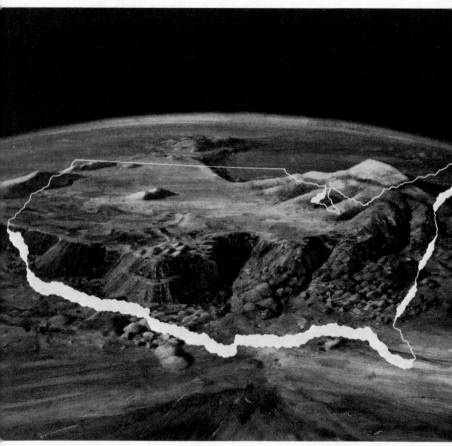

Artist's conception of the Ishtar Terra highland region of Venus, always hidden beneath the thick clouds. The continental United States is outlined to scale. Near the Great Lakes rises mammoth Maxwell Montes. Courtesy NASA.

Alps. If Mont Blanc were set in the lowest depression of Venus, it would rise only 6,300 feet (1,920 meters) above plain level, and Maxwell Montes would rise 28,000 feet (8,841 meters) higher—almost as high as the Earth's great Everest.

Maxwell Montes

In the eastern Ishtar Terra highlands on the planet Venus rises one of the greatest mountains in the solar system, Maxwell Montes. Towering as high as 35,300 feet (11,800 meters) above the so-called "plain level" of Venus (an imaginary sphere with the planet's average radius of 3,758 miles, or 6,051.4 kilometers), it is almost 1.2 miles (2 kilometers) higher than Earth's Mount Everest.

Radar measurements from Earth-based instruments and from the orbiting Pioneer-Venus radar system depict the Maxwell mountain region as extremely rough terrain, the roughest on Venus. Radar brightness suggests that Maxwell's steep slopes are covered with rocks larger than 2.5 inches (6.3 centimeters). Pioneer-Venus has also discovered a dark feature on the eastern flank of Maxwell Montes, with a diameter of about 60 miles (100 kilometers) and a depth of more than 3,000 feet (1,000 meters). It is believed to be a volcanic crater some 3 times larger than the largest caldera on planet Earth. Over 9 New York Cities could fit at the bottom of this huge cone crater, and even the taller tower of the World Trade Center, with a height of 1,350 feet (412 meters), would be buried over 1,600 feet (487 meters) below the crater's rim. Since Maxwell's summit is some 6.68 miles (10.7 kilometers) high, it is likely that there are hundreds of lightning bolts cracking around it sometimes—the altitude at which they occur.

MARS

Land, Land Everywhere

Mars is a much smaller world than the Earth, even though its rotation period and axis tilt make it more Earthlike than any other planet. With a diameter of 4,222 miles (6,794 kilometers), it is a little more than half the Earth's diameter of 7,926 miles (12,753 kilometers). Even with its much smaller size, however, the red planet has larger mountains, canyons, polar caps, and craters than the Earth. But Mars has no oceans, no bodies of water, and this fact gives Mars a land area almost equal to that of the Earth's—56 million square miles (145 million square kilometers) compared with the Earth's 57.5 million square miles (149 million square kilometers)—much of which becomes enveloped in a great dust storm every Martian spring.

The Long Months of Mars

A day on the planet Mars is just slightly longer than an Earth day—24 hours, 37 minutes. The tilt of its rotation axis, about 25 degrees from the orbital plane around the Sun, is also very similar to the Earth's 23.5-degree tilt. The days and seasons of Mars are therefore very similar to our planet, even though a Mars year, the time it takes it to orbit the Sun, is almost twice that of Earth: One Mars year equals 669.6 Mars days.

If a Mars calendar were set up using our twelve months as a base, each Mars month would be about 55 days long, and

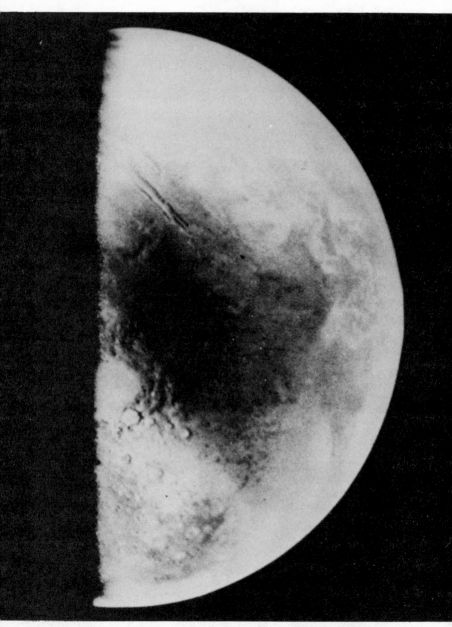

Mars   Viking 1 approach to Mars shows more detail than any Earth-based telescope—a first look at the new Mars. Courtesy NASA.

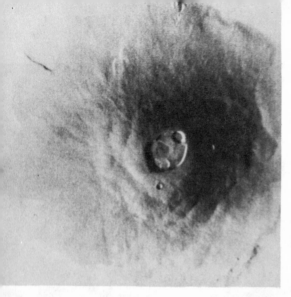

Olympus Mons, the largest known volcano in the solar system. It is over two-and-a-half times the height of Mount Everest and its base would cover the state of Missouri.

The flanks of Olympus wreathed in clouds, with its huge summit caldera rising above them.

Rare closeup of the caldera, with small landslide visible. Courtesy NASA.

there would be dates like March 43 and September 51—much more time to pay those end-of-the-month bills.

The Colors of Mars

The color of the daylight sky on Mars, when no dust storms are raging, has been described as salmon pink—a color that would only be seen in some sunset skies on Earth. This occurs because air motions caused by temperature changes stir the dust on the surface and carry it into the atmosphere. It is the scattering of sunlight on these iron-rich dust particles, the same that give the surface its orange-red color, that give the red planet its pink daytime sky. During a Mars sunrise or sunset, however, the sky is deep or dark blue, similar to what astronauts see high above the Earth, because convection currents are inactive at these times and there is little dust aloft. Nighttime skies on Mars are crystal clear, black sprinkled with colored stars—an observational astronomer's dream—with no bothersome atmospheric twinkling.

Olympus Mons

The largest known volcano in the solar system is Olympus Mons on the planet Mars. It rises about 79,000 feet (24,000 meters) above the Martian desert. The Earth's highest, Mount Everest, is only 29,028 feet (8,850 meters) above sea level. The base of Olympus Mons is 370 miles (600 kilometers) across and would cover the state of Missouri. The volcanic crater at the summit is over 40 miles (64 kilometers) across, large enough to accommodate Anchorage, Alaska, *and* New York City. If Edmund Hillary set out to climb Olympus Mons at the same pace as he climbed Everest, it would take him 130 days—almost three times as long as Everest.

Valles Marineris

Mars is a small world of giant geological features. Besides the largest known mountain in the solar system, the red planet also has the largest canyon known in the solar system. Named Valles Marineris (Valley of Mariner) after the Mariner 9 space probe that discovered this vast interconnected canyon system during its Martian reconnaissance in 1971–72, this gigantic rift stretches one sixth of the way around Mars. Located near the planet's equator, its west end is on the summit of Tharsis Ridge, a highland region where many of the great Martian volcanoes are located. Some planetary scientists think that these canyonlands were formed by two crustal plates separating.

Valles Marineris is some 2,800 miles (4,500 kilometers) long, 370 miles (600 kilometers) wide in some places, and 4.5 miles (7 kilometers) deep. It is 13 times longer than the Grand Canyon, the Earth's largest land gorge, which would fit into it as a small tributary. If Valles Marineris were superimposed to scale on a map of North America, it would stretch from New York to California; it would cut the continent in half if it were filled with water.

The Great Martian Landslides

Mariner and Viking orbital photography showed many great landslides along the canyon floors of Valles Marineris. The largest of them is 62 miles (100 kilometers) long—a mass of Martian terrain that slid into a deep Marineris canyon millions of years ago. If such a slide occurred on the eastern seaboard of North America, it would wipe out the coast from New York City to Toms River, New Jersey, more than half the distance to Philadelphia.

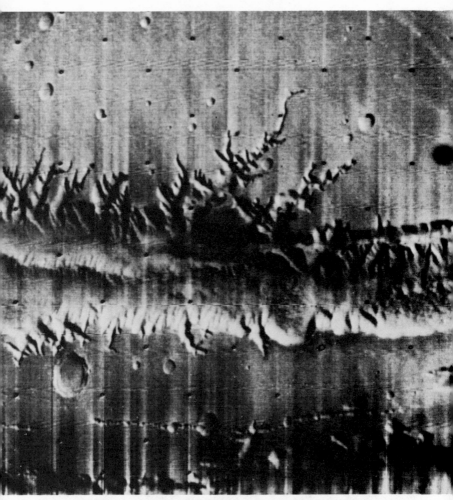

A view of Valles Marineris, the "Grand Canyon" of Mars, which could stretch from New York to California. The Earth's Grand Canyon could fit into it as a small tributary. Courtesy NASA.

A Martian landslide. The largest ones are over 60 miles (100 kilometers) in length and on Earth would wipe out half the coastline from New York to Philadelphia. Courtesy NASA.

Hellas of Mars

The southern hemisphere of Mars is geologically old—perhaps 3.5 billion years—and strewn with impact craters, while the northern half of the planet has a much younger surface, since great volcanoes have given it a new face with their vast lava flows.

The southern plateau contains two large impact basins—Argyre and Hellas, the latter being the largest impact crater on Mars and one of the largest known in the solar system. Hellas is 1,000 miles (1,600 kilometers) across and 3.1 miles (5 kilometers) deep. It is one of the brightest areas on the planet and can even be observed from Earth through a telescope under ideal conditions. Planetary scientists speculate that the brightness in this giant basin may be caused by frost or clouds or dust. Observations show that major dust storms start in or near Hellas, which have often obscured the basin floor. The steep slopes of Hellas provide the topography for vigorous winds, making the Hellas basin not just the solar system's largest known impact crater but also its largest known dustbowl. Hellas could swallow up the eastern seaboard of the United States from Boston, Massachusetts, to Jacksonville, Florida, and conceal it in a Martian dust storm.

Hellas Formed

The great Hellas basin on Mars was excavated by an asteroid with a diameter of at least 100 miles (161 kilometers). It blasted into the Martian surface with a velocity of about 36,000 miles (58,000 kilometers) an hour and produced an explosion equal to 100 trillion tons of TNT. Just 1 ton of TNT will completely destroy a large apartment building or a small city block of buildings.

An Inhospitable Atmosphere

The thin Martian atmosphere is composed of 95 percent carbon dioxide, 2.7 percent nitrogen, 1.6 percent argon, and traces of several other gases, including 0.15 percent oxygen. At the Viking lander sites, surface pressure was less than 1 percent of the Earth's surface pressure, but it does vary by season because most of the atmosphere becomes frozen at the planet's polar caps during the winters. To experience an atmosphere like that of Mars on Earth, a person would have to climb to an altitude of about 20 miles (32 kilometers). A jet plane flying at about 100,000 feet (30,500 meters) would encounter an atmospheric pressure similar to the surface of Mars. If an astronaut stepped from his spacecraft onto the Martian surface without protection, he would suffocate from lack of oxygen after only half a minute or so—even faster than he would freeze from the cold.

Outlook: Cold and Deadly

Venus is too hot, Earth is (much of the time) just right, and Mars is too cold for people. Temperatures on Mars vary widely between day and night, winter and summer, mainly because it is about 50 million miles (80 million kilometers) farther from the Sun than the Earth. At the Viking lander 1 site in the northern hemisphere, summer temperatures ranged from a low of −88 degrees Celsius (−126 degrees Fahrenheit) just before dawn to a high of −12 degrees Celsius (−10 degrees Fahrenheit) in the midafternoon. Polar temperatures during the winter drop as low as −140 degrees Celsius (−220 degrees Fahrenheit), and equatorial temperatures at noonday may briefly reach the human comfort level of 20 degrees Celsuis (68 degrees Fahrenheit). Even at the comfortable noonday equator, however, the intense ultraviolet radiation from the Sun would easily penetrate the thin atmosphere and eventually kill all living things on the surface. Human skin would be severely sunburned in just 15 minutes.

The Waters of Mars

Most of the water on Mars is concentrated in the polar caps and in the permafrost. The wettest atmosphere, found in the far north during the summer, is more than 100 times drier than the air we breathe on Earth. There is so little water vapor in the Martian atmosphere that if all of it were condensed, it would cover the planet with a film no more than one thousandth of an inch (a few hundredths of a millimeter) thick. If collected in one place, it would form a body of water about the size of Walden Pond. But if the water in the polar caps were melted, it could form a planet-wide ocean about 33 feet (10 meters) deep. By contrast, if all Earth's water were spread evenly over the surface, it would flood our world to a depth of 9,000 feet (2,700 meters). The total amount of water on Mars is enough to flood an ocean over twice the land area of the United States to a depth of 300 feet (91 meters).

Where the Rivers Once Surged

There are ancient dry river channels on Mars; the largest are about 620 miles (1,000 kilometers) long and over 100 miles (161 kilometers) wide. They were probably formed by sudden gigantic surges of water, far more intense than any flash flood. Geologists have compared these Martian channels with the ''channeled scab lands'' of Washington and Idaho in the United States. Thousands of years ago in these states a glacier formed a dam of ice that blocked the outflow of water from other melting glaciers. The ice dam finally gave way, and water surged in a tremendous flow that was 10 times greater than all the rivers on the Earth combined. In just a few days this powerful flow eroded and cut deeply into the basalt of the channeled scab lands—no doubt similar to the Martian surges.

The Martian Clouds

There are two kinds of clouds on Mars: clouds composed of water or water ice, such as those on Earth, and those composed of carbon dioxide. The dense carbon-dioxide clouds form over the northern polar cap in the Martian autumn as the cap condenses from the rich carbon-dioxide atmosphere. This cloud cover is called the *polar hood* and hides from view the waxing northern polar cap.

The clouds of water-ice crystals often resemble our own cirrus clouds and form downwind of a large mountain or crater. They are much higher in the atmosphere than cirrus clouds are over Earth, however—about 60,000 feet (18,293 meters) above the Martian surface. Water-ice or water clouds also form in Martian cyclones and other storm fronts in the late fall and winter. Viking lander 2 detected the passage of these storms every 3 to 4 days. The Martian cyclones occur at relatively low altitudes—about 4 miles (6.4 kilometers) above the planet—and are formed when cold polar air flows under the low, warmer air. One of the few cyclonic storms observed on Mars during the half decade of Viking was about 375 miles (600 kilometers) across. This storm was about 80 percent the size of the largest hurricane ever recorded on Earth, so there is little doubt that Mars holds the record for the largest cyclonic storm—even if we have never seen it.

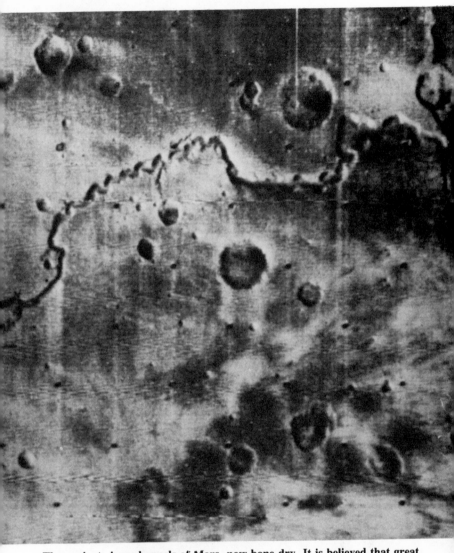

The ancient river channels of Mars, now bone dry. It is believed that great surges of water once flowed, 10 times greater than all the rivers of the Earth combined. Courtesy NASA.

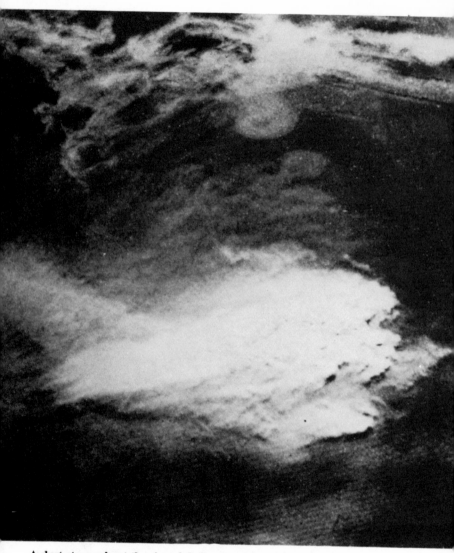

A dust storm about the size of Colorado in the western portion of Valles Marineris. Courtesy NASA.

Capping Off Mars

The polar caps of Mars wax and wane with the seasons, not by melting but instead by evaporating in the summer and condensing in the winter. As the caps grow during the winter, carbon-dioxide clouds cover them as more and more of the atmosphere freezes out and becomes the solid dry ice of the poles. A full 20 percent of the carbon-dioxide atmosphere of Mars freezes into dry ice each season.

There remains a large permanent northern polar cap during the Martian summer, and there is strong evidence that it is composed mostly of water ice, which even in summer remains way below the freezing point. The residual south polar cap is reduced to a much smaller area than the north cap, and it apparently has no water ice, just dry ice.

If there were an increase in the Sun's radiation, the Martian ice caps would melt and dramatically change the red planet. The water ice would melt and evaporate, and as more water vapor entered the atmosphere, it would create a greenhouse effect and prevent the escape of heat from the planet. The atmospheric temperature would rise, liquid water would condense, and the melting caps would form shallow seas. This once-cold desert planet would have clouds and rain and rivers. It would be the beginning of a new Earth.

Phobos: The Senile Moon

Phobos, the larger of the two tiny Martian moons (17 by 13.5 by 12 miles; 27 by 21.5 by 19 kilometers), is one of the darkest bodies yet observed in the solar system because its surface has extremely low reflectivity. Its future looks equally dark. Because Phobos orbits only 3,718 miles (5,982 kilometers) above Mars, tidal effects are causing its orbit to decay, and it will plunge into the Martian surface in about 100 million years, blasting out an impact crater over 60 miles (100 kilometers) in diameter. While 100 million years is a great span of time compared to an average human life, it is just a tick in the life of our solar system, which is about 4.5 billion years old. Thus Phobos has already lived 98 percent of its life. Its orbits are numbered.

Deimos: The Tiny Moon

The second moon of Mars, Deimos, is even smaller than Phobos, its dimensions being 6.8 by 7.4 by 9.3 miles (11 by 12 by 15 kilometers). Like its potato-shaped companion, it may be a captured asteroid lured by Mars's gravity at some time in the past. Deimos is so small and its gravity so weak that people could launch themselves into space by merely running along its surface and attaining the modest speed of 7 miles (11 kilometers) per hour—the escape velocity for this tiny moon of Mars.

JUPITER

Sizing Up the Giant

Jupiter the giant, well known as the largest planet in our solar system, contains 71 percent of the total mass of all the planets. Its diameter of 88,748 miles (142,796 kilometers) is 11 times that of the Earth, but a comparison of surface areas is much more impressive. If the surface of Earth could be peeled off like an orange skin and spread out against the surface of Jupiter, its area would be to Jupiter as India is to Earth. Left

The melting North polar ice cap at its smallest; dark areas are devoid of ice, probably due to wind patterns. Courtesy NASA.

Jupiter, the giant, seen by Voyager 1 at a distance of 20 million miles (33 million kilometers). If the Earth were represented by a dime, then Jupiter would be a dinner plate. Courtesy NASA.

unpeeled, the Earth would be the size of a dime when Jupiter was the size of a dinner plate.

Spin, Jupiter, Spin

Jupiter is the fastest rotating planet in our solar system, spinning on its axis once almost every 10 hours at its equator. A stationary object at its equator would be traveling at 27,720 miles (44,601 kilometers) per hour, which is substantially more than the escape velocity from planet Earth—the speed at which a spaceship launched from Earth can escape from our planet's gravitational shackles and into the depths of space. Voyager 1 almost reached this speed when it was launched September 5, 1977, toward Jupiter, and it took almost 1,300 spins of Jupiter (each Jupiter day about 10 hours long) to reach the giant planet, compared to about 540 days that passed on Earth.

The Giant's Magnetosphere

A compass needle on giant Jupiter would point to the south, not to the north, because its magnetic field is reversed compared to our Earth's. Jupiter's magnetic field, in fact, the most powerful of any planet in the solar system, has a total magnitude about 18,000 times the Earth's and dominates the region of space around the planet. This entire region is Jupiter's magnetosphere, one of the largest structures in our solar system, its shock-wave boundary formed by its interaction with the solar wind—the Sun's outward flow of charged particles that travel at 0.9 million miles (1.45 million kilometers) per hour. This immense magnetosphere has a diameter of about 12.4 million miles (20 million kilometers), within which a large flattened disk of electrically charged particles exists; the entire magnetosphere and its disk rotate around the planet. If Jupiter's magnetosphere radiated in a frequency that was visible to the human eye, it would appear twice the size of the full Moon from Earth. If one edge of this magnetosphere met the Earth's atmosphere, it would extend about 52 times farther than the Moon. Jupiter's inner magnetosphere is so intense that if an Apollo spacecraft flew through it, the astronauts would be killed in minutes.

Almost a Star

The composition of Jupiter is very similar to that of our own Sun; 89 percent hydrogen and 11 percent helium. Its average density is only 1.3 times that of water; it is a gaseous sphere with no solid surface. Jupiter radiates about twice as much heat energy as it receives from the Sun, which comes from an internal reservoir of heat left over from its creation some 4.6 billion years ago. This thermal heat slowly works its way up to the surface and, along with the rapid rotation, helps to maintain a remarkably constant temperature between the equator and the poles and between the day and night hemispheres—about −138 degrees Celsius (−216 degrees Fahrenheit).

Great Jupiter is about as large as a planet can be without being a star. If it had more mass, it would not grow in size as might be expected; rather, the extra mass would cause it to shrink from self-compression under gravity. Indeed, if Jupiter were about 60 times more massive than it is, thermonuclear reactions would ignite, and the energy produced would overcome the gravitational shrinkage. Jupiter would then become a self-luminous star instead of a planet, with a diameter of more than 100,000 miles (161,000 kilometers).

Jupiter's Core

The deep interior core of Jupiter, probably composed of iron and silicates and about the size of the Earth or Venus, has a temperature of about 30,000 degrees Celsius (54,000 degrees Fahrenheit). While this is 100 times hotter than any terrestrial surface, it nevertheless is 500 times cooler than the temperature at the center of the Sun. The tremendous pressure in Jupiter's core is more than 30 million times higher than the Earth's atmosphere. If planet Earth were ever subjected to such immense pressures, it would be compressed to little more than half its present diameter and to a density 3 to 4 times that of iron.

The Jovian Sea

Surrounding Jupiter's core is a shell of metallic hydrogen over 25,000 miles (40,000 kilometers) thick, an exotic form of hydrogen never found on Earth. Although we usually think of hydrogen as a gas, under the high pressures in Jupiter's interior, it is compressed into a metal. At the outer limit of this metallic hydrogen shell, the temperature has cooled off to 10,500 degrees Celsius (19,000 degrees Fahrenheit) and the pressure has dropped to about 3 million atmospheres. An outer shell, about 15,000 miles (24,000 kilometers) thick, is composed of liquid hydrogen, and this continues until the hydrogen becomes gas, which marks the depth at which Jupiter's interior becomes atmosphere and cloud layers. In this sense, Jupiter's entire "surface" is a sea of liquid hydrogen at a temperature of at least −252 degrees Celsius (−422 degrees Fahrenheit).

A Saturn V Launched from Jupiter

Jupiter has the highest escape velocity of any planet in the solar system—136,371 miles (219,600 kilometers) per hour—because its huge mass gives it the strongest gravity.

Would a Saturn V moon rocket escape from Jupiter's powerful gravity if launched from the planet's strange liquid-hydrogen "surface"? At first the rocket would sit on the launch pad until it had burned off about half the propellant in the first stage. Then it would struggle upward and eventually reach a speed of about 2,500 miles (4,000 kilometers) per hour and cover a range of only 30 miles (48 kilometers). The second stage would then start burning, but before this stage burned up enough fuel where its thrust could be effective against Jupiter's gravity, it would have crashed. So the same rocket that could deliver 50 tons to the Moon from Earth would do no better than a long-range artillery shell if launched from Jupiter's surface.

The Great Red Spot

A storm has been raging on Jupiter for over 300 years, a storm that could swallow up the Earth—three of them, in fact. The great red spot, a vast cyclonic system, has had a diameter as large as 25,000 miles (40,000 kilometers), 3 times the diameter of the Earth. Fifty of the Earth's largest hurricanes could be placed side by side before they matched the width of this largest known storm in the solar system.

A Many-Splendored Atmosphere

The complex and swirling atmosphere of Jupiter, with its ever-changing multicolored cloud currents, is about 600 miles

(1,000 kilometers) thick. Not even the technically sophisticated cameras of the Voyager spacecraft could see beneath the upper portions of this opaque cover, composed mainly of ammonia crystals.

Above and below Jupiter's equatorial region there are light- and dark-colored bands that run parallel to the equator. They result from Jupiter's rapid rotation, which smears these cloud patterns into parallel currents. The bright zones, usually white or yellow, represent the colder and higher clouds, where the gases are ascending, driven upward by the convection of warm gases. The dark belts are the denser, warmer, and lower regions of the atmosphere, which are generally descending. These reddish-brown belts are believed to be the lower circulating clouds. During its close approach, Voyager 1 was targeted to photograph a large brown-colored oval, a common feature in Jupiter's northern latitudes, because scientists believed it represented an opening in the upper cloud deck. These brown oval features have a lifetime of one or two years, while the more turbulent regions near features such as the great red spot are relatively short-lived, a fact borne out by the time interval between photographs taken by Voyagers 1 and 2. The great red spot is an exception to the color-altitude rule. It is a high-atmospheric feature despite its color, probably because convection brings up traces of phosphorus from the lower atmosphere.

Wind speeds in Jupiter's atmosphere vary at different latitudes, and generally the wavy, swirling patterns have slower velocities than the more smeared streamlike patterns. An example of the latter, a pale-orange stream called the north temperate current, has constant wind speeds of about 260 miles (418 kilometers) per hour, over twice the velocity of severe hurricane winds on Earth. Similar high atmospheric currents, along with cyclonic storms such as the great red spot and the white oval, represent Jupiter's "weather," which only occurs in the top 50 miles (80 kilometers) of the giant's thick 600-

mile (1,000-kilometer) atmosphere. Below, on the global ocean of hot liquid hydrogen, all is calm.

The Jovian Clouds

The high white and yellow cloud tops of Jupiter are composed of frozen ammonia crystals and have temperatures in the range of −130 degrees Celsius (−202 degrees Fahrenheit). A lower cloud layer is composed of ammonium hydrosulfide crystals at about −73 degrees Celsius (−100 degrees Fahrenheit). The lowest cloud layer lies about 43 miles (69 kilometers) below the tops of the high, visible ammonia clouds and 560 miles (900 kilometers) above the liquid hydrogen ocean. Temperatures would be in the 30-degrees-Celsius (86-degrees-Fahrenheit) range. Here, clouds of water droplets and ice crystals are found. At somewhat lower altitudes, the warmer temperatures would create steam. These water regions of the Jovian atmosphere are more Earthlike than any other feature of the planet.

Exactly how Earthlike? Could, for example, a future astronaut on a Jovian atmospheric expedition open up a portal of his ship and let the outside in? The answer is no. The atmosphere would still be mostly hydrogen and helium, without a trace of breathable oxygen. In addition, it would contain quantities of such noxious, smelly gases as ammonia and hydrogen sulfide, topped off with doses of phosgene and cyanide.

Jupiter's Outermost Moon

Sinope, the tiny outermost moon of Jupiter's family, was discovered in 1914 by astronomer Seth Nicholson. That it was discovered at all is amazing, considering that its diameter is not much over 9 miles (15 kilometers) and it is almost 400 million miles (640 million kilometers) away from Earth.

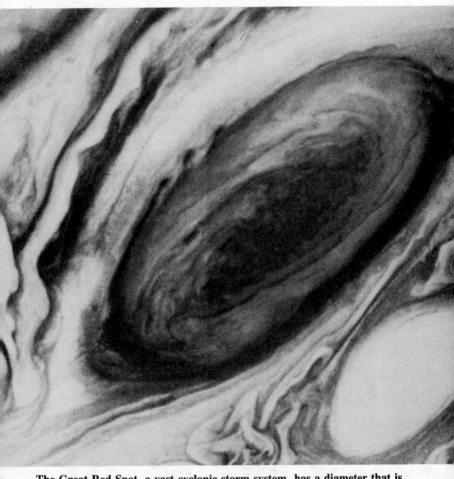

The Great Red Spot, a vast cyclonic storm system, has a diameter that is three times larger than the Earth's. Fifty of the world's largest hurricanes could be placed side by side within it. Courtesy NASA.

Jupiter and its four big moons are seen in this composite photograph. From background to foreground, they are: Io, Europa, Ganymede, and Callisto. Courtesy NASA.

Sinope is about 14.7 million miles (23.6 million kilometers) from Jupiter, and it takes more than 2 years (758 days) to orbit the planet. It is therefore 61 times farther away from Jupiter than our Moon is from Earth, and about one sixth the Sun–Earth distance. An Apollo spacecraft traveling at its average Earth–Moon velocity would take half a year to travel from Jupiter to its outermost moon, Sinope.

Io, the Active

Io, comparable in size to our Moon, was one of the four large inner satellites of Jupiter discovered by Galileo in 1610. The Voyager spacecraft missions in the late 1970s sent back astonishing close-up views of this amazing world—which had been no more than a bright spot in Galileo's crude 30-power refractor telescope—and discovered eight active volcanoes as well as extensive sulfur deposits on its surface, resembling those of the Utah salt flats.

Io is so close to Jupiter, 261,000 miles (421,000 kilometers)—slightly farther than the Moon is from Earth—that Jupiter raises huge tides in its solid surface, as the Moon raises tides in Earth's oceans. Other nearby large Jovian satellites force Io to rock from side to side, and the tidal bulge then shifts position. These shifts heat the satellite's interior, like an auto tire heats by being flexed as it rolls along the pavement. This heating has melted Io's core, producing vast amounts of lava which erupt continually. Little Io has the most active volcanism in the solar system and is therefore the most geologically active body known. Because of the vast lava flows, Io's surface is less than 10 million years old. It is the only body in the solar system that is turning itself inside out volcanically. On Earth, the agent of renewal is not volcanoes but earthworms.

The Giant's Rings

As the Voyager 1 spacecraft moved inside the orbit of Amalthea, one of the smallest and innermost moons of Jupiter, it discovered a ring of debris around the planet which remains invisible from Earth because of its thinness and the overwhelming brightness of Jupiter. It is the third ring system to be discovered in our solar system, Saturn's being the first and Uranus's the second.

Its outer edge is about 34,000 miles (55,000 kilometers) above Jupiter's cloud tops, and its width is estimated at about 3,700 miles (5,920 kilometers). This thin ring, about 0.6 of a mile (1 kilometer) thick, has two segments: a bright segment about 500 miles (800 kilometers) wide and a dim segment 3,200 miles (5,120 kilometers) wide. While the inside structure of the ring has not been resolved because of the limits of Voyager's cameras, it is believed that particle size ranges from microscopic to many feet (a few meters) in diameter. Jupiter's ring owes its existence to the recently discovered small moon 1979 J3, which is about 25 miles (40 kilometers) in diameter and whose weak gravity acts to prevent particles from leaving the ring. This satellite orbits just outside the ring's outer edge, almost like a marble rolling around on the outside of a hoop. If all the debris from the rings were compressed into a solid body, it would be only slightly larger than Jupiter's odd-shaped, tiny moon Amalthea, with a diameter of 93 miles (150 kilometers), which is almost 1,800 times less than the rings' orbital diameter.

Europa's Global Ice Age

Europa is the smallest of the four large moons of Jupiter discovered by Galileo in 1610. With a diameter of 1,925 miles (3,100 kilometers), it is slightly smaller than the Earth's

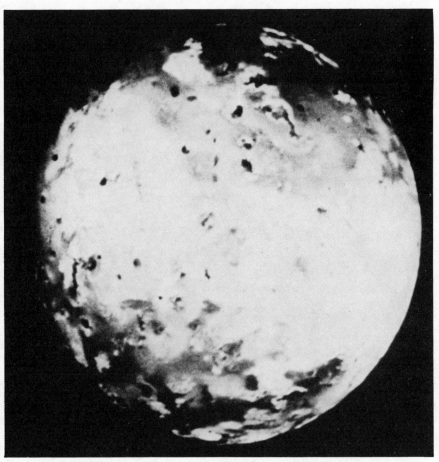

Over 100 volcanic features were identified on Io, including 8 active volcanos. As a result, Io's surface has vast snow fields of sulfur. Courtesy NASA.

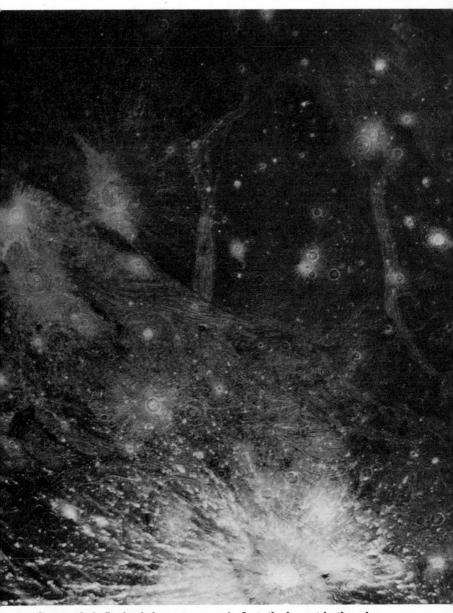

Ganymede is Jupiter's largest moon—in fact, the largest in the solar system. Half of its mass is composed of water or ice. The grooves on its icy surface are unique; they are evidence of tectonic movement, which is rare on moons. Courtesy NASA.

Moon, which is 2,159 miles (3,476 kilometers) in diameter.

The surface of Europa is probably a thin crust of ice, at a temperature of about −130 degrees Celsius (−202 degrees Fahrenheit), with slush 60 miles (100 kilometers) thick underneath. This highly reflective crust makes it the brightest Galilean surface. It also has one of the smoothest surfaces of any moon in the solar system; nowhere are there hills or other features as high as 300 feet (100 meters), and no craters more than 3 miles (5 kilometers) in diameter were observed by the Voyager spacecraft. Its most prominent features are a number of long grooves, fractures in the icy crust, that resemble the famous canals of Mars once thought to exist. These linear fractures are over 600 miles (1,000 kilometers) long in some places and 125 to 185 miles (200 to 300 kilometers) thick. They may be remnants of deep cracks in the ice that were originally 100 miles (160 kilometers) or more in depth but have since been filled with material from beneath, which most likely results from internal tidal flexing which heats the thin outer ice crust. Some planetary scientists predict that there is a deep ocean of liquid water beneath the ice shell, which would resemble the Arctic or the Ross Ice Shelf on Earth. What may swim in such an ocean is anybody's guess.

Ice World

Galileo's eyes were no doubt first drawn to bright Ganymede as he peered through his telescope in 1610 and discovered Jupiter's four large moons. Ganymede is Jupiter's brightest and largest moon; it orbits beyond Io and Europa, taking just over a week to revolve around the planet. It is about 1½ times the size of our Moon or some 3,167 miles (5,100 kilometers) in diameter—larger, in fact, than the planet Mercury.

Ganymede is believed to have a thick outer layer of ice,

and Voyager photographs show fresh-looking craters that appear like chips in a frozen lake made by a hammer. The striking hammers, of course, were large meteoroids that blasted into the surface, leaving their impact craters. One huge impact basin, about 2,000 miles (3,200 kilometers) in diameter, remains from the ancient crust, much of which has been erased by glacial movements over the icy surface. This basin is comparable to the great impact basins of the Moon and Mercury.

Besides the heavily cratered regions, Ganymede has a curiously grooved terrain of closely spaced parallel ridges and troughs, most about 10 miles (16 kilometers) wide and several hundred long. These grooves are unique to Ganymede in the solar system and are the first certain examples of lateral faulting to be found on a moon or planet other than Earth. These grooves are evidence of tectonic movement in Ganymede's icy crust, and they suggest that this moon's crust was once very active, similar in certain ways to the plate tectonics of Earth.

Ganymede is about half as dense as the Earth's Moon, or 1.9 times the density of water, because fully half its mass is composed of water or ice. Heat Ganymede up, and more than half of it would evaporate.

Cratered Callisto

Callisto, the most distant major moon of Jupiter, is almost 1.2 million miles (1.9 million kilometers) away and orbits once every 16.7 days. It too is an ice world, just slightly smaller than Ganymede, with the lowest density of Jupiter's four large moons. This probably means that it is composed of large amounts of ice and water. The surface of Callisto is darker than the other Galilean moons but still twice as bright as the surface of our Moon. Its terrain is also the oldest of the four big moons; large areas probably date back to the era of intense meteoritic bombardment ending about 4 billion years ago, shortly after the formation of the solar system. Callisto's claim to fame is that it has more craters than any other planetary body in our solar system.

The Fastest Moon

The fastest-moving moon in the solar system, discovered in the Voyager spacecraft photos and officially designated 1979 J3, swings around Jupiter once every 7 hours 4.5 minutes. Only 25 miles (40 kilometers) in diameter, it revolves around the giant at an average distance of over 35,000 miles (56,549 kilometers) above the Jovian cloud tops. Its velocity has been calculated at 70,400 miles (113,600 kilometers) per hour—a speed that would take you from New York to San Francisco in 2 minutes 11 seconds.

SATURN

Buoyant Saturn

Saturn is composed mostly of hydrogen and helium, like its giant neighbor, Jupiter, and has a density that is lower than water (0.69 of the value 1.0 for water), the lowest of any planet in our solar system.

If a chunk of Saturn could be brought to Earth—more than a 900-million-mile (1,448-million-kilometer) journey—at a celebration party in honor of the event, the piece of low-density planet could float in the punch bowl.

Saturn's Three-Decade Year

Saturn revolves around the Sun once every 29.5 Earth years, making a Saturn year about three decades long. The planet travels in its orbit at a speed of about 21,600 miles (34,754 kilometers) per hour. The ringed planet thus orbits around the Sun just 2.4 times during the average human life span.

Saturn Inside Out

Saturn is, like Jupiter, a giant sphere of gas, mostly hydrogen and helium, with no solid surface. Deep in its interior, however, it is believed to have an iron and rocky core, estimated to be about 17,000 miles (27,000 kilometers) in diameter—twice the size of the Earth or almost ⅐ Saturn's sphere. This core is so compressed that its mass is 15 to 20 times the mass of the Earth. Around it is a thick ice layer, and surrounding the ice and core layers is a thick shell of metallic hydrogen, something never observed on Earth because of the immense pressures needed to form it. Liquid hydrogen and helium form the thickest outermost layer of the gaseous sphere; it is their interaction that probably causes Saturn's high energy output. Above this "surface" is the outer atmosphere, consisting mostly of hydrogen, helium, and methane.

High above in the cloud tops, the winds of Saturn blow with twice the velocity of those on Jupiter—some 900 miles (1,400 kilometers) per hour. On Earth this would level all the great cities.

The Subtle Clouds of Saturn

The cloud tops of Saturn are 40 degrees colder than those of Jupiter—a bitter −170 degrees Celsius (−274 degrees

Fahrenheit)—because of the planet's greater distance from the Sun. As a result of these colder temperatures, ammonia condenses into clouds at lower altitudes in Saturn's atmosphere, and some planetary scientists argue that this process creates a haze that obscures the cloud patterns below. Whatever the reason, there are few dramatic and vivid markings in Saturn's atmosphere as compared to Jupiter. From Earth, Saturn's cloud cover usually appears calm, subtly colored with pastels of yellow and orange. Cold temperatures probably inhibit outbursts in cloud activity that produces the great red spot and white ovals of Jupiter.

There was one well-known exception to Saturn's generally calm appearance, however. In August 1933 a great white spot was observed on the equator which expanded over a few days until the entire equatorial region of Saturn brightened. This white spot was estimated to be 20,000 miles (32,000 kilometers) long and 12,000 miles (19,000 kilometers) wide—the largest temporal feature ever observed on the ringed planet, rivaling Jupiter's great red spot. The reddish-brown oval feature that Voyager 1 discovered in Saturn's southern hemisphere was only about one third this size. The 1933 white spot could have engulfed an object more than twice the size of the Earth. Almost 7 planet Mercurys could fit sphere to sphere along its length.

Saturn, the Generator

Like Jupiter, Saturn has an internal source of energy and generates more than twice as much energy as it receives from the Sun. Comparing planetary masses, this makes Saturn's energy output even more intense than Jupiter's.

Where does this energy come from? A probable explanation is that Saturn is colder than Jupiter, and therefore its interior behaves differently. It appears that, deep within Saturn, helium is separating out from the hydrogen in which it is mixed, and the helium is seeping downward to form a core. Since helium is heavier than hydrogen, this would release the observed amount of energy. How much energy does Saturn generate? About 130 trillion kilowatts, equal to over 100 million large electric generating plants.

Snowstorm in Orbit

Second in size to Jupiter and sixth out from the Sun at an average distance of about 892 million miles (1,435 million kilometers), the planet Saturn is considered by many to be the most beautiful planet in our solar system. Saturn's rings do their slow angular dance about every 15 years, presenting their different aspects to observers on Earth during this period.

When backlit by the Sun, parts of the rings resemble frosted colored glass. They are composed of trillions of particles, each one a tiny separate satellite in its own orbit. Estimates of their size have varied widely during the past few years; radar reflections from the rings obtained in 1972 suggested that the particles have diameters between 1.5 inches (4 centimeters) and 1 foot (30 centimeters). No doubt they come in all sizes, from dust-sized specks to boulder-sized chunks, depending on which ring they orbit in. They are composed of water ice and their surfaces appear fluffy like snow, so the rings may be likened to a permanent snowstorm in orbit about Saturn.

Saturn Up Close

From a vantage point slightly above the rings, a person would see the Sun as a brilliant point of light, only one tenth as large as the Sun we see from Earth. The rings would be a

Beautiful Saturn as seen by Voyager 1 from a distance of 50 million miles (81 million kilometers). Because of the planet's rapid rotation, its poles are substantially flattened. Courtesy NASA.

The rings, looking like a celestial phonograph record, are composed of water and ice and come in all sizes, from dust-sized specks to boulder-sized chunks. Before Voyagers 1 and 2, astronomers knew of 6 separate rings. As Voyager 1 approached, the number grew: first 100, then 500, finally as many as 1 thousand individual ringlets were counted. Courtesy Jet Propulsion Laboratory, NASA.

bright flat expanse extending to the horizon, cutting the universe sharply with a plane so geometrically flat as to appear entirely artificial. One could fly tens of thousands of miles and still see that clean, flat plane cutting the world sharply in two. Saturn would fill the view like a movie screen when we sit near the front of the theater. It would span more than 100 times the width of the full Moon, showing bands of clouds amid its hazy yellow color and rotating almost as we watched—once every 10 hours and 39 minutes at the equator.

Rings of Light

The rings of Saturn are more show than substance. The ice-composed or ice-covered particles reflect most of the sunlight—some 70 percent—and it is this fact that makes Saturn's rings dazzling to the observer's eye over great distances. At certain times and positions, in fact, the rings are actually brighter than the planet.

The mass of the rings is modest in comparison with their vast extent—probably equal to that of a large comet, but still 1 million times less massive than our Moon and about 82 million times less massive than the Earth. The rings are believed to be made mostly of ice, but they probably contain less ice than in the Earth's polar regions.

The Ring Alphabet

The outermost bright ring of Saturn, designated ring A, is the second brightest, and its farthest edge extends 47,500 miles (76,500 kilometers) from the planet's cloud tops. Then there is a 1,615-mile (2,600-kilometer) gap, named the Cassini division after the astronomer who discovered it in 1675. This gap contains thin rings or bands of material, about twenty, which resemble the contrails of high-flying jet aircraft in Voyager photos. Next in, ring B is the widest and brightest and leads into the fainter ring C, also known as the crepe ring. Then comes ring D, discovered just in 1969; it is the most faint and innermost ring, confirmed by Voyager, and probably extends almost to Saturn's cloud tops.

Saturn's rings are probably much less than 1 mile (1.6 kilometers) thick, even though this value is still debated. Assuming that they are 0.621 mile (1 kilometer) thick, then comparing this thickness with a square section of the rings, one side equal to the distance from the outermost edge of ring A to Saturn's cloud tops, some 47,500 miles (76,500 kilometers), is like imagining a sheet of paper whose dimensions are 30 square feet, as big as a living room.

Swift Passage

Saturn's rings stretch out for 170,000 miles (274,000 kilometers) at their maximum diameter. The Voyager and Pioneer spacecraft passed just under the planet's rings during their close encounters and thus were influenced by Saturn's powerful gravity field. These spacecraft traversed the full extent of the ring system in about 4 hours, traveling at almost 43,000 miles (69,000 kilometers) per hour. At this velocity, a rocket would take just 6 hours to travel from the Earth to the Moon or only 35 minutes to orbit the Earth once.

Sweeper of the Ring, Mimas

While several dark divisions between the rings of Saturn have been discovered in recent years, only the well-known Cassini division discovered in 1675 by the Italian astronomer is clearly visible through a small telescope from Earth. This

One of Saturn's many
surprises, discovered by
Voyager: the radial spokes
of the B ring that persisted
and perplexed scientists.
Courtesy NASA.

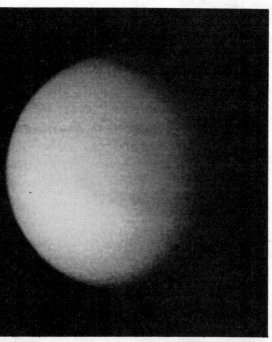

Two large moons of
Saturn: Giant Titan, with
its thick nitrogen
atmosphere and cratered
Mimas which partly
sweeps clear the Cassini
division. Courtesy Jet
Propulsion Laboratory and
NASA.

dark division is between the two brightest rings, A and B, both of which are quite dense with icy ring particles, especially A.

Why does the Cassini division exist? It is there because of Saturn's small moon, Mimas, which orbits outside ring A and partly sweeps this wide zone clear by displacing many of the ring particles (that would otherwise orbit there) with its weak gravity. But the gap is 2,600 miles (4,183 kilometers) wide, and it has been difficult to explain why little Mimas (218 miles or 351 kilometers in diameter) creates so large a gap. The reason is that Saturn's rings behave like a galaxy with tightly wound spiral arms. The gravity of the rings acts to amplify the weak gravitational effects caused by Mimas. If Jupiter's satellites may be likened to a miniature solar system, then Saturn's rings resemble a miniature galaxy—at a scale of 1 to 3 trillion. On this scale, if the Galaxy is the size of the Earth, Saturn's rings are the size of a bacillus.

Titan: The Giant Red Moon

Titan is the largest of Saturn's fifteen-plus moons as well as the second largest moon in the solar system. Titan's diameter of 3,094 miles (4,978 kilometers) makes it larger than the planet Mercury and three quarters the size of Mars. But Jupiter's Ganymede is larger by 80 miles (129 kilometers). Besides its near-record size, it is the only satellite in the solar system to retain a substantial atmosphere, composed mostly of nitrogen. Reddish smog cloaks its cold surface, which has a prevailing temperature of about −186 degrees Celsius (−303 degrees Fahrenheit). The atmospheric smog is produced by chemical reactions.

Many astronomy and space artists have sought to render beautiful Saturn as it would appear from Titan's surface. But because of the thick smog surrounding this giant red moon, confirmed by Voyager 1, a visitor would not see Saturn at all.

Target Titan

On November 11, 1980, the Voyager 1 spacecraft passed Saturn's moon Titan at a distance of only 2,500 miles (4,000 kilometers) from the surface—the closest approach to any celestial body encountered by either of the two Voyagers. At this close fly-by, it was more than 946 million miles (1,524 million kilometers) from Earth and traveling at a speed of about 43,000 miles (69,000 kilometers) per hour. Such cosmic target accuracy is comparable to shooting an arrow at an apple 6 miles (9.6 kilometers) away and having the arrow pass the apple at a carefully calculated distance of 1 inch (2.54 centimeters).

Celestial Snowballs

The other moons of Saturn are all much smaller than planet-sized Titan. Iapetus and Rhea have diameters of about 894 and 950 miles (1,440 and 1,530 kilometers). For Tethys and Dione, the diameters are around 650 miles (1,045 kilometers). Other Saturnian satellites—including Mimas, Enceladus, Hyperion, and Phoebe—have diameters in the 50- to 325-mile (80- to 523-kilometer) range.

The densities of Saturn's moons are all quite low. Titan's, the best determined, is only 1.32 times that of water, so all these satellites are mostly ice. They may be the largest snowballs we will ever see.

The Moon Mimas

Saturn-hugging Mimas, orbiting about 115,000 miles (186,000 kilometers) from the planet's center, has a heavily cratered surface. Voyager photos showed that one face of this moon was dominated by a huge crater about 80 miles (130 kilometers) in diameter, a full third of the moon's 242-mile (390-kilometer) diameter. The crater walls rise an estimated 5.6 miles (9 kilometers) high—possibly the highest in the entire solar system. This means that little Mimas has crater walls rising 29,510 feet (8,997 meters), higher even than Mount Everest.

Curious Iapetus

Iapetus is one of the outermost moons of Saturn, orbiting about 2.2 million miles (3.5 million kilometers) from the planet's center. Its diameter is some 894 miles (1,440 kilometers). That plus the fact that Voyager flew past it at a distance of about 5 million miles (8 million kilometers) gave scientists a fuzzy and indistinct image. Even at this distance, however, Iapetus's most curious feature was obvious: its light and dark faces. One hemisphere is about 6 times brighter than the other, a contrast unique in our solar system, and scientists have only theories at present to explain it. The most plausible one is that the dark material reached the surface as a result of internal eruptions. What is sure, though, is that Iapetus proves that many mysteries of Saturn and moons remain, even after the revelations of Voyager.

The Giant Montage

The Voyager 1 and 2 spacecrafts took more than 70,000 television pictures of Jupiter, Saturn, their satellites, and the space environment around them. If all these photographs were developed in an 8-by-10-inch standard format and positioned side by side to form a rectangular area, the sides would measure 8.82 miles (14.2 kilometers) by 11.03 miles (17.76 kilometers), with a total area of 97.5 square miles (252 square kilometers), about equal to twice the area of Boston, Massachusetts.

URANUS

Roll-on Uranus

The most remarkable feature of Uranus is that it is tilted on its axis more than any other planet. While Earth is tilted from its orbital plane by some 23.5 degrees, Uranus's tilt is nearly 98 degrees. At times during its orbital motion, the north or south pole is aligned nearly face on toward the Sun, so the planet actually rolls along its orbit. During those times the poleward hemisphere receives nearly constant sunlight, while the other hemisphere languishes in decades-long darkness. This is responsible for creating the solar system's longest seasons, winters and summers that are 21 years long. In 1985, Earth observers will look directly at the north pole of Uranus as its winter rolls around before them.

Green Uranus

The atmosphere of the planet Uranus is mostly hydrogen and helium, like those of Jupiter and Saturn, but it also appears to contain 0.4 percent of methane. Since methane strongly absorbs red light, the reflected sunlight from the planet shows mostly blue, green, and yellow light. As a result, Uranus appears green—the emerald planet of our solar system.

The Dark Rings of Uranus

In 1977 rings were discovered around Uranus—at least nine in all. They are all very thin and quite similar to the single ring around Jupiter, discovered in 1979 by Voyager 1. The Uranus rings range from 26,090 to 31,873 miles (41,979 to 51,284 kilometers) out from the planet's center, the most inner ring being about 10,000 miles (16,000 kilometers) above the cloud tops. Six of the rings are no more than 3 to 6 miles (5 to 10 kilometers) wide, and the 3 widest have widths of 10, 30, and 60 miles (16, 48, and 97 kilometers). They are therefore quite unlike the rings of Saturn, with their widths of tens of thousands of miles. Like Jupiter's ring, the ring material is probably held in place by small adjacent satellites, but they have not yet been observed. This means that there are several new moons to be discovered around Uranus.

The rings of Uranus are composed of dark material which reflects less light than blackboard slate. This is why it took an airborne telescope, flying aboard a modified C-141 aircraft at an altitude of almost 8 miles (12.5 kilometers), to discover these faint rings that are about 1.7 billion miles (2.7 billion kilometers) away from Earth.

The Hard Uranian Winter

The longest winter (that is, when the Sun is far south of the celestial equator) in the solar system is 21 years long and occurs on the south polar regions of Uranus. The winter temperature is estimated to be −219 Celsius (−362 degrees Fahrenheit), so that future astronauts exploring the liquid hydrogen "surface" will need much more than thermal underwear to keep them warm.

Distant Pictures

Uranus has five known satellites. Outward from the planet, they are: Miranda, Ariel, Umbriel, Titania, and Oberon. These moons form a very regular system, whose orbits are nearly circular and in the same plane. The largest, Titania, is about 1,100 miles (1,800 kilometers) in diameter; the smallest, Miranda, is about 340 miles (550 kilometers) in diameter. Our first good look at them, and at Uranus itself, may come early in 1986 when Voyager 2 flies by.

When Voyager 2 gets to Uranus, it will be some 2 billion miles (3.2 billion kilometers) from planet Earth. This is nearly 100 times farther than the 1960 record for deep-space communications set by Pioneer 5, which communicated from 22 million miles (35 million kilometers) out, but did not send image data: If all goes well, Voyager's encounter with Uranus and its moons will show an improvement in space communications, since 1960, by a factor of about 1 million.

☆   ☆   ☆   ☆   ☆
NEPTUNE

Distant Twins

Neptune is the farthest "giant" planet from the Sun, orbiting beyond Jupiter, Saturn, and Uranus at an average distance of almost 2.8 billion miles (4.5 billion kilometers). This puts the planet a full 1 billion miles (1.6 billion kilometers) farther away than its "sister" giant, Uranus—almost 11 times the Sun–Earth distance, 4,167 times the Moon–Earth distance, and 288,267 times the distance between New York and Lon-

A computer-simulated print, depicting Voyager 2 flying by Uranus and its newly discovered rings in 1986. Courtesy NASA.

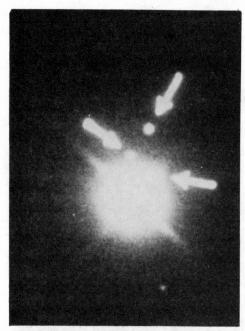

Uranus and its three largest moons as seen through Earth-based telescopes. Voyager 2 may give a closer look in 1986. Lick Observatory photo.

Neptune and its giant moon Triton as seen from Earth. For the next 20 years, Neptune is farther away from Earth than Pluto. Lick Observatory photo.

don. If a converted Boeing 747 could be made space-worthy, flying at its maximum speed of over 600 miles (965 kilometers) per hour it would take 1,903 years to travel from Uranus to Neptune.

A Neptune Year

No person on planet Earth has ever lived to be 1 Neptune year old. A Neptune year is the time it takes the planet Neptune to revolve once around the Sun—164.8 Earth years. Every person now living on Earth will be long dead when the planet Neptune returns to its present position in orbit. But there will be, we hope, plenty of our great-great-great-great-grandchildren alive.

Neptune's Temporal Claim

For 20 years beginning in 1979, Neptune—not Pluto—is the most distant planet from the Sun. This is because Pluto's orbit carries it inside Neptune's orbit for a 20-year period once every 248 years. All science textbooks will have to be revised or at least footnoted to account for Neptune's temporal claim. Neptune held the planetary distance record for 84 years before Pluto's discovery, and now we know that it will never fully relinquish it.

Triton's Planetary Tree

Neptune has two moons, Triton and Nereid, and both have unusual motions, the causes of which are not understood.

Triton is one of the largest satellites in the solar system, larger than the Earth's Moon but substantially smaller than Saturn's giant Titan. It is large enough to retain a thin atmosphere of methane that was discovered in the 1970s. Triton's diameter is 2,360 miles (3,800 kilometers), and a circular orbit puts it at an average distance of 219,800 miles (354,000 kilometers) above Neptune, closer than our Moon is to Earth.

What makes Triton so unusual, however, is that its circular orbit is inclined 20 degrees to the planet's equator and that its motion is opposite to the direction of Neptune's rotation—in other words, Triton orbits backward once every 5 days 21 hours and it is the only large moon in the solar system to do so. This peculiar orbit, as well as that of Neptune's other moon, Nereid, may have resulted from interaction with another moon ejected from the system in the distant past. Some astronomers speculate that this escaped moon became the small planet Pluto.

Eccentric Nereid

Nereid, the small outer moon of Neptune, is probably no larger than 200 miles (322 kilometers) across. Its highly eccentric orbit, more so than any other moon in the solar system, takes it from about 870,000 miles (1,395,000 kilometers) to over 6 million miles (9.6 million kilometers) from Neptune.

Nereid's peculiar orbit, along with companion Triton's backward motion around Neptune, points to an unusual event in Neptune's past—an event that may have involved a close encounter between Triton and Pluto. Whatever the cause, Nereid's eccentric orbit resembles a comet's more than a moon's, and it may eventually allow this tiny Neptunian moon to escape and go wandering in the outer solar system.

PLUTO

Faint, Faraway Pluto

Pluto, aptly named for the god of the dark underworld, orbits in Stygian darkness at an average distance of almost 3.7 billion miles (6 billion kilometers) from the Sun. From Pluto's surface, our Sun would appear no larger than the planet Jupiter does from Earth at its closest point. The faint reflected sunlight from Pluto's surface, 1,000 times fainter than Neptune's, takes about 5 hours and 40 minutes to reach the Earth. Even Voyager 1, traveling constantly at its rapid Saturn-encounter speed of 45,255 miles (72,815 kilometers) per hour, would take 81,468 hours (9.3 years) to cover this distance.

Shrinking Pluto

Pluto is the smallest and lightest planet in our solar system, more like a methane snowball than like the inner planets Mercury, Venus, Earth, and Mars, all of which are solid and rocky. Indeed, many astronomers think that Pluto should be considered as the largest and most distant asteroid rather than a full-fledged planet, as it has been labeled.

Before Pluto was discovered by Clyde Tombaugh in 1930, astronomers had predicted the existence of a planet beyond Neptune, based on small irregular motions in the orbits of Uranus and Neptune. Percival Lowell began a dedicated search for this trans-Neptunian planet in the first decade of this century, but nothing was found. He predicted that planet X would have a mass 7 times that of Earth. A fellow astronomer, William H. Pickering, calculated that the undiscovered planet's mass was 2 times the Earth's. Predictions of its diameter were equally large, up to several times the size of the Earth.

Ever since its discovery in 1930, estimates of Pluto's size and mass have been shrinking. The first values used were larger than Earth—about 10,000 miles (16,000 kilometers) in diameter. In the 1960s, estimates of Pluto's diameter dropped its size to about one half that of Earth, or some 3,700 miles (6,000 kilometers). In the late 1970s, the discovery of Pluto's moon, Charon, finally brought Pluto down to size, making it the smallest planet in our solar system. It is now known to be about 1,800 miles (2,900 kilometers) in diameter, with a mass of only one four-hundredth of the Earth or one fifth that of our Moon. Pluto's mass turns out to be 14,000 times less than that predicted for planet X by Percival Lowell at the beginning of the twentieth century. Little Pluto, smaller than many moons of the solar system, is too small to be planet X, the hypothetical planet that was believed responsible for slight irregularities in the orbits of Uranus and Neptune. Instead, Pluto remains, a half century after its discovery, the most enigmatic and distant body beyond the Earth.

Oddball Pluto

One of the main reasons that some astronomers believe Pluto should be stripped of its planet status and reclassified a minor planet or a giant cometary body is its peculiar orbital path around the Sun, different from any other planet and the most eccentric. All the other planets circle the Sun in about the same plane, but not Pluto—it is inclined by more than 17 degrees to the plane of the other orbiting planets. Pluto also has the most elongated orbit, which is an exaggerated ellipse unlike the more gentle ellipses of the other planets. This odd orbit takes Pluto as far as 4,582 billion miles (7,375 billion kilometers) from the Sun and as close as 2,750 billion miles (4,425 billion kilometers) to it.

If our Earth's orbit around the Sun had the same elliptically shaped orbit as Pluto, it would approach to within 12 million miles (19 million kilometers) of Mars at its farthest point from the Sun and to within a mere 2 million miles (3.2 million kilometers) of Venus at its closest approach—only about 8 times farther than our Moon.

The Greatest Gulf

The greatest distance between two planets in our solar system is between Pluto and Neptune, which occurs once every 496 years. Their maximum separation is 7,403 billion miles (11.91 billion kilometers)—nearly 80 times the Sun–Earth distance or 1 million times the diameter of the Earth.

A New Moon for Pluto

Charon, Pluto's moon, was discovered by James W. Christy of the U.S. Naval Observatory in 1978. Like its planet, Charon was found by a careful study of photographic plates. It was seen as no more than an undefined bump on the extremely fuzzy photographic image of Pluto.

The new moon's diameter is estimated to be about 435 miles (700 kilometers), and it is separated from Pluto by about 12,500 miles (20,000 kilometers). This makes Charon the largest moon compared to its planet in the entire solar system—a distinction that our own Earth–Moon system held until 1978. Charon is also the second closest moon to a planet, runner-up to Mars's Phobos with a distance separation of only 5,828 miles (9,378 kilometers).

Pluto and Charon are locked together by their gravity and keep the same sides facing one another. Because of this fact and their close size relationship, they are like a double planet. Their rotation period is the most accurately known feature of the hard-to-observe Pluto system—6 days 9 hours 16 minutes 54 seconds. A Plutonian day is therefore almost as long as a week is on Earth. Since Pluto's year equals 247.7 Earth years, this means that there are over 90,000 of our days in every Pluto year (a year equal to more than 3 average human lifetimes) without moon Charon, locked in place, ever rising or setting over the icy landscape.

The Tricky Light of Pluto

If it had not been for Pluto's surface of methane ice or frost, discovered by analyzing the spectra of the planet's reflected sunlight, it might not have been found as early as 1930. This is because the ice or frost makes tiny Pluto extremely bright, since the surface reflects rather than absorbs most of the sunlight. This surface brightness is one of the reasons that Pluto was considered much larger than it actually was for almost 50 years.

Pluto's surface, the coldest of any planet, is too cold for an atmosphere. The estimated average surface temperature of -230 degrees Celsius (-382 degrees Fahrenheit) would freeze out any atmosphere and keep it on the surface. While analysis of the reflected sunlight indicates that the planet may contain some solid matter such as silicates, this is difficult to reconcile with Pluto's density, determined from the size and mass that were calculated after discovery and study of the new moon, Charon. Pluto's density is close to that of water, with a possible low value of 0.6 that of water. If the lower-than-water density proves correct, a chunk of icy Pluto would bob in the Atlantic Ocean before evaporating.

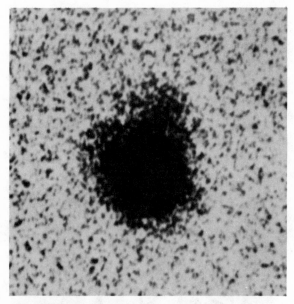

The discovery plate of Pluto's recently discovered
moon, Charon. The fuzzy bulge at the top is Charon.
This image has been enlarged 100 times, and so the
film grains are brought out. Courtesy James W.
Christy and the U.S. Naval Observatory.

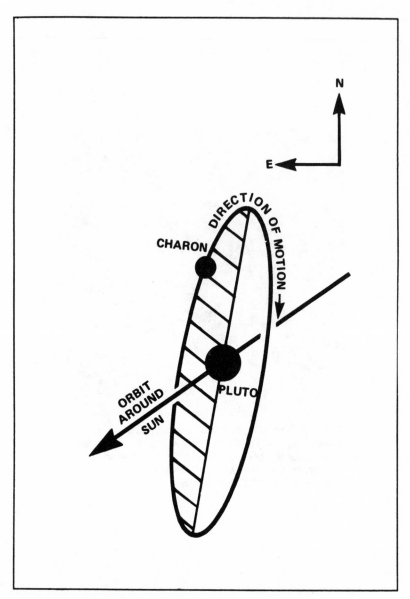

The probable orbit of Charon around Pluto, with approximate sizes to scale. Courtesy James W. Christy and the U.S. Naval Observatory.

Pluto's brightness varies by 20 percent during its rotation period every 6.4 days, but this is not due to the moon, Charon. Nor is the brightness change gradual; rather, it increases slowly and diminishes quickly. This implies an irregular surface with dark spots, but astronomers must wait for sophisticated telescopes in Earth orbit before this can be confirmed.

If Pluto and Charon were in the Moon's orbit around the Earth, they would be 3 times brighter than the Moon because of all the reflecting ice. But the bright nights on Earth would not last long, because these two methane ice worlds would evaporate in the sunlight. They would be shrouded in thick atmospheres produced from the evaporating gases; the gases of Charon would escape to form a ring around Pluto. At a somewhat slower rate, the gases of Pluto would also escape, forming a ring around Earth. These rings would continually be replenished as more ice evaporated on Charon and Pluto, even though the solar wind would blow away some of this gas. The rings would be especially visible at night, appearing luminous, like brilliant comet tails.

THE MOON: FULL, HALF, AND IN BETWEEN

☆ ☆ ☆ ☆ ☆

MOON BASICS

Lunar Safari

The surface of the Moon has about the same area as Africa. Unlike Africa, however, the Moon's surface shows the tracks of only one kind of creature.

The Earth's Partner

Our Moon is the fifth largest moon in the solar system, surpassed in size only by Jupiter's Ganymede (the largest) and Callisto, Saturn's Titan, and Neptune's Triton. The diameter of the Moon is about one fourth the size of the Earth—2,160 versus 7,927 miles (3,475 versus 12,755 kilometers)—and it most closely resembles the size of Jupiter's Io, which has a diameter of 2,310 miles (3,716 kilometers), as well as the most active volcanoes in the solar system. In relation to the size of its parent planet, the Moon took first place until 1978, when Pluto's moon, Charon, was discovered. Now Charon holds the title. Our Moon's mass is one eightieth that of the Earth, while Triton, Neptune's Moon, has a mass ratio to its planet of 1 to 750. Charon's mass ratio to Pluto, however, is 1 to 10—the largest known in the solar system. Indeed, the Moon and Charon are so much larger proportionately to their planets than all the other moons that some planetary scientists view both systems as double planets. The Moon's large size relative to our planet is responsible for the strong tidal motions which have probably influenced the evolution of life on Earth.

The Moon's Muscle

The Moon causes tidal motions not just in the Earth's oceans and other bodies of water but also in its atmosphere and solid body. The lunar and solar tidal effect in the center of the ocean is between 2 and 3 feet (0.6 to 0.9 meter), but this height varies greatly and often becomes much higher, depending on water depth, shoreline shape, and other factors. The Earth's atmospheric tide results in a slightly increased air pressure. When the Moon is directly overhead, it causes a solid Earth tide that raises up North America, or any other land mass, about 6 inches (15 centimeters). So everyone is walking tall at least twice a month—during new and full moon.

Maintaining a Comfortable Distance

If the Moon were closer to the Earth—50,000 miles (80,000 kilometers) away rather than its mean distance of about 239,000 miles (385,000 kilometers)—the tidal forces would be tremendously more powerful and would flood the coastal regions of the world under hundreds of feet of water.

New York City would have to move to the high ground of the Adirondack Mountains, and Paris would have to go to the Ardennes Plateau in northeastern France.

Smooth Seas and Dry Weather

The Moon is 1 million times drier than the Gobi Desert, and the only floods on the Moon have been floods of molten hot lava that poured over vast areas of the Moon's surface to create the smooth-looking maria (''seas'') that have been seen with the naked eye for millions of years.

A Rarefied Atmosphere

The Moon has an atmosphere but an extremely tenuous one, which would be considered a high vacuum by terrestrial standards. Apollo instruments detected a collisionless gas composed of helium, neon, argon, and radon, most of which probably comes from the solar wind. If all the molecules in 1 cubic centimeter (0.061 cubic inch) of this extremely thin Moon ''atmosphere'' were lined up end to end, they would fit into the period of this sentence. If you did the same thing to breathable Earth atmosphere, the lined-up molecules would go to the Moon and back—almost half a million miles—with some to spare.

The Two-Faced Moon

The Moon's visible face is puffy but smooth; it is puffy because there is a pronounced bulge on this side, possibly because of the strong tidal influences of the Earth; it is smooth because almost half the near side's face is covered with maria—great expanses of ancient lava flows that spread over the floors of huge impact basins. The Moon's far face is thick-skinned and pockmarked: thick-skinned because it has a much thicker crust than the visible face (62 miles or 100 kilometers on the far side, 37 miles or 60 kilometers on the near side), and pockmarked because there was no extensive lava flooding as on the near face, and therefore most of the ancient crater impact scars remain since the Moon's formation. Almost all the dark maria material formed by ancient lava flows is on the near side of the Moon. (One of the very few exceptions to this is the crater Tsiolkovsky on the far side, which was filled up with the darker lava.) This means that the Earthside face was the favored side for volcanism—probably because of the influence of greater tidal forces—and so the near face of the Moon got a face lift about 3 billion years ago when the great lava floods came.

Light Math

Logic, not lunacy, would agree with this statement: The full Moon is twice as bright as the half-moon. But this is incorrect. The fact is that a full Moon is nine times as bright as a half-moon, since the surface of the visible half-moon is extremely rough and mountainous, which makes for more shadows and less reflected sunlight. As the Moon waxes toward fullness, the sunlight falls on more smooth reflecting surface, and brightness increases proportionately. Einstein knew it: Light can play tricks on us.

Moon at Noon

As bright as the full Moon appears on a cloudless and clear night, the fact is that it is only reflecting about 7 percent of the sunlight that falls on its surface. The other 93 percent is absorbed by the surface and accounts for noontime temperatures at the lunar equator of over 100 degrees Celsius (212 degrees Fahrenheit), the boiling point of water. Had one of the Apollo astronauts dropped an egg into a small declivity on a lunar boulder, it would have dropped slowly in the one sixth of Earth's gravity, without risk of breaking, and would have boiled vigorously and congealed quickly in the hot sunlight and vacuum of the Moon's surface.

The Moon's Look-alike

Except for the crater and maria patterns, the Earth 4 billion years ago probably looked very much like the Moon does today. Then something happened.

Different Worlds

Much of the Moon's surface was molten rock soon after it formed from the solar nebula about 4.6 billion years ago. As this molten layer began to cool, huge asteroids began their intense bombardment and continued for millions of years. Then, at about the 4-billion-year mark, the cataclysmic impacts became less frequent and eventually died away. At the same time, the interior was heating up slowly because of the decay of radioactive elements, and for the next half-billion years, about 3.8 to 3.1 billion years ago, the Moon's interior burst forth with great floods of lava that pooled in the large impact basins to create the maria (''seas'') that are seen today. As the last lava floods flowed, the Earth was active with mountain-building, great oceans, running rivers, and the stirrings of life.

The Great Lunar Floods

The Moon's great lava floods continued for 800 million years between 3 and 4 billion years ago. The fluid lava basalt spread out across the floors of the ancient craters for hundreds of thousands of square miles. The individual eruptions were rapid because of the fluid nature of the lavas, which flowed like heavy engine oil at room temperature and spread out widely in thin sheets. When it flowed over the Imbrium Basin, one of the largest features of the Moon, with a diameter of about 1,000 miles (1,600 kilometers), the thin sheets spread out for almost 800,000 square miles (2 million square kilometers), an area equivalent to one third the size of Australia.

The Rain of Rock and Dust

The cosmic rain of meteorites and micrometeorites onto the lunar surface is unrelenting and has been going on for billions of years. It has been estimated that the material falling onto the Moon's surface amounts to about 4.9 feet (1.5 meters) every billion years. This material is mixed with the pulverized material created during the explosive meteorite impacts, the largest of which occurs on the average of about once every 100 million years. The seismic equipment left on the Moon by the Apollo crews has measured between 70 and 150 meteorite impacts a year, caused by meteorites weighing between 3.5 ounces (100 grams) and 2,205 pounds (1,000 kilograms). The largest recorded thus far weighed about 1 ton and impacted on the far side of the Moon in July 1972.

Gardening on the Moon

The lunar soil (regolith) is constantly being turned over by meteorites and micrometeorites rather than by spades and hoes. This is called "gardening," but it more closely resembles a heavily bombarded battlefield than a freshly dug vegetable garden. Every million years an entire thin surface layer of the Moon's soil is turned over 100 times. Every billion years almost one half inch (1 centimeter) of lunar topsoil will be turned over at least once. Geologic time on the Moon is slow-motion compared to the Earth—except when a large meteorite hits.

The Writhing Moon

The crater Giordano Bruno is believed to be one of the youngest craters on the Moon, blasted out from the lunar surface only 800 years ago. But more important, recent research indicates that the cosmic impact event was actually witnessed by five or more men in southern England. The odds against an impact event of this magnitude occurring on the Moon in recorded history is 1 in 1,000 (for a 3,000-year period). The odds against an eyewitness account are even more astronomical, but the June 18, 1178, event was written in the medieval chronicles by Gervase of Canterbury. The men were facing the new Moon when "suddenly the upper horn split in two . . . [and] a flaming torch sprang up, spewing out, over a considerable distance, fire, hot coals, and sparks. Meanwhile the body of the moon which was below writhed, as it were, in anxiety, and . . . throbbed like a wounded snake." After careful analysis, scientists concluded this recorded event was the crater Giordano Bruno in the making—a large crater with a di-

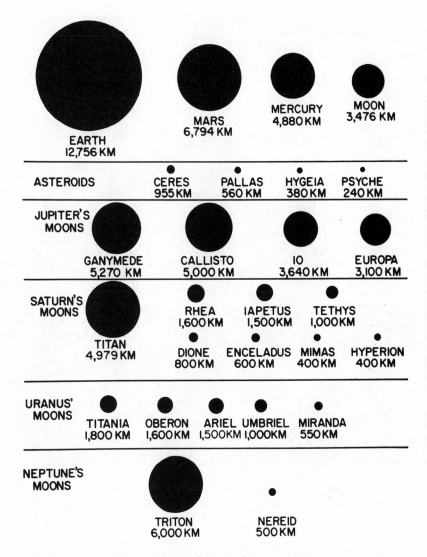

EARTH
12,756 KM

MARS
6,794 KM

MERCURY
4,880 KM

MOON
3,476 KM

ASTEROIDS

CERES
955 KM

PALLAS
560 KM

HYGEIA
380 KM

PSYCHE
240 KM

JUPITER'S MOONS

GANYMEDE
5,270 KM

CALLISTO
5,000 KM

IO
3,640 KM

EUROPA
3,100 KM

SATURN'S MOONS

TITAN
4,979 KM

RHEA
1,600 KM

IAPETUS
1,500 KM

TETHYS
1,000 KM

DIONE
800 KM

ENCELADUS
600 KM

MIMAS
400 KM

HYPERION
400 KM

URANUS' MOONS

TITANIA
1,800 KM

OBERON
1,600 KM

ARIEL
1,500 KM

UMBRIEL
1,000 KM

MIRANDA
550 KM

NEPTUNE'S MOONS

TRITON
6,000 KM

NEREID
500 KM

Our Moon compared to the planets Earth, Mars, and Mercury as well as the other satellites of the solar system. Courtesy Science Graphics, copyright © 1982.

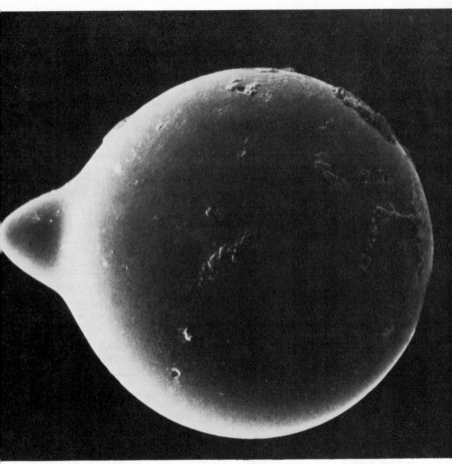

Electron microscopic image of tiny emerald-green glass droplet with tail. Many were found in soil returned from Spur Crater area by Apollo 17 crew. Courtesy National Space Science Data Center.

The meteoroid blast that formed the bright-rayed crater Giordano Bruno
was witnessed by several people in southern England 800 years ago.
Courtesy NASA.

Charles Duke of Apollo 16, near the rim of North Ray Crater, which may
have been formed by a comet strike. Courtesy NASA.

ameter of about 12.5 miles (20 kilometers) that has a bright
and fresh-looking ray pattern spreading out hundreds of miles
from it.

Counting Craters

About 500,000 craters on the Moon can be seen from the
Earth through the largest and most powerful telescopes. It
would take a person over 400 continuous hours to count them
all—and these don't even include the craters on the Moon's
opposite side.

The Largest Crater

Bailly, a great walled plain near the southern limb of the
Moon, is 183 miles (294 kilometers) in diameter and is the
largest named crater on the visible side of the Moon. Because
it is an ancient formation, billions of years of cosmic erosion
by meteorites and micrometeorites have broken and worn
down some of its features to such a degree that it has been
called a "field of ruins." Nevertheless, the mountain peaks of
its walls still rise to 14,000 feet in many places. Only space
flight and space-age technology have given us a true picture of
its stature as a lunar feature, because its position near the south
pole of the Moon distorts our optical view of it. Bailly is not a
deep crater (about 2.5 miles; 4 kilometers), and its floor—
which contains a large and well-formed crater B—is light in
color. While there are larger craters on the far side of the
Moon—Hertzsprung, for instance—and larger maria features
on the near side that were originally crater formations, no one

can dispute Bailly's greatness. Its square area of about 26,000 square miles (67,300 square kilometers) could contain the states of West Virginia and Rhode Island, as well as the cities of Boston, Chicago, Omaha, Albuquerque, and Los Angeles. The lunar Rover vehicle would take a week at top speed to travel around Bailly's circumference, if it were possible.

Greater Craters

Two of the greatest (and the youngest) lunar basins were excavated about 4 billion years ago within a few hundred million years of one another: the Imbrium Basin and the Orientale Basin. Then for a period of about 600 million years the great lava flows flooded their floors (much more in Imbrium than Orientale) and they became lunar seas (maria): Mare Imbrium and Mare Orientale. Both maria have been called the Moon's largest craters, casting aside the traditional claim of Bailly. The fact that both have been traditionally designated maria and did experience lava flooding is adequate reason to define them as maria rather than craters, although no one will dispute the fact that they once were mammoth impact craters.

Mare Orientale is younger than Imbrium and is famous for its impressive three-ring mountain chain that has been interpreted by some as being successive shock waves of lava that rapidly cooled and froze solid. Orientale is about 560 miles (900 kilometers) in diameter and was blasted out by an asteroid that was 15 miles (25 kilometers) in diameter. Mare Imbrium is older, larger, and flooded with more lava. It is almost 1,000 miles (1,600 kilometers) in diameter, and the massive asteroid that shook the entire Moon was about 93 miles (150 kilometers) in diameter. This so-called Imbrium event was so powerful that it flung material across the entire visible face of the Moon. It was equal to 100 billion of the largest hydrogen bombs.

The blasting out of the Imbrium Basin, caused by an impacting asteroid about 4 billion years ago. The explosion was so powerful that the debris was flung across the entire face of the Moon. Courtesy U.S. Geological Survey.

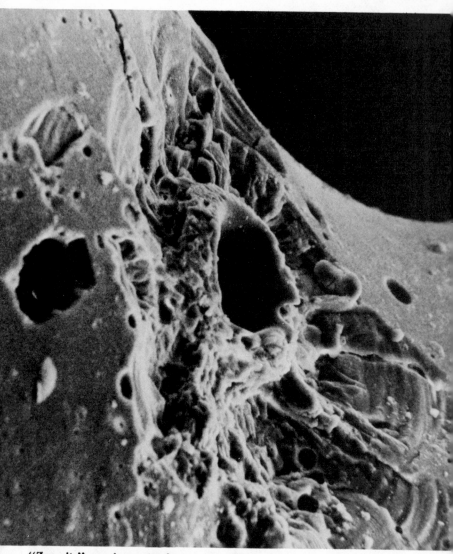

"Zap pits" or micrometeorite craters, visible only through an electron microscope, are constantly formed in lunar material and create lunar soil, "regolith," over time. Courtesy David S. McKay and NASA.

A Trail of Cosmic Bullets

Since there are literally billions of small microcraters ("zap pits") that cover the entire surface of the Moon—every square inch—it would be impossible to find the smallest without putting the whole lunar surface through an electron microscope. Tiny grains of rock and metal called micrometeorites continually strike the lunar surface, creating microcraters. Over millions and billions of years, this unending bombardment has created the lunar soil (regolith) and rounded off the crater rims. Most micrometeorites are from $\frac{1}{10,000}$ to $\frac{1}{1,000}$ of an inch in size and travel at about 70,000 miles (113,000 kilometers) per hour. These cosmic bullets pack quite a wallop, possessing 100 times more energy than an equivalent mass of TNT. They created the powdery soil in which was left the first human footprints on the Moon.

An Impressive Crater

Crater Copernicus, considered by many to be the most impressive crater on the lunar surface, has been nicknamed "the Monarch of the Moon." Copernicus, with its extensive ray system, dominates the Oceanus Procellarum just west of the lunar central line and is prominent at full Moon. This crater's massive walls reach heights of 17,000 feet (5,183 meters), twice as high as the Grand Canyon, and its diameter, measured from crest to crest, is 56 miles (91 kilometers). Ejected material from Copernicus was found by the Apollo 12 astronauts near their landing site, about 250 miles (400 kilometers) away from the crater. Copernicus was formed (a euphemism for blasted out) about 1 billion years ago when an asteroid-sized rock about 1.25 miles (2 kilometers) in diameter struck the lunar surface with a force equivalent to a million one-megaton hydrogen bombs.

The Moon's Brightest Spot

The crater Aristarchus was named for the Greek mathematician who determined that the distance to the Sun was 28 times longer than the Moon's distance from Earth. It is by far the brightest formation on the visible side of the Moon. When the famous astronomer Sir William Herschel viewed it in 1787, he thought it was an active volcano in eruption. This crater is 25 miles (40 kilometers) in diameter, 2.25 miles (3.6 kilometers) deep, has a central peak and terraced walls, and is relatively young—about 200 million years old. The age and bright ray patterns of Aristarchus are not the only reasons for its amazing brightness. More than half the so-called Transient Lunar Phenomena (TLPs) sighted involve this crater, and Armstrong, Aldrin, and Collins, the Apollo 11 crew, reported a luminous north wall during their historic first-landing mission to the Moon. Later, during the flight of Apollo 15 in 1971, gaseous emissions were measured and confirmed. Radon gas was detected, which indicates the area is radioactive. So Aristarchus is not just bright; it also glows. It is an active place and no doubt has some more surprises in store.

The Moon's Uplands

The rough and bright areas of the Moon are known as highlands or uplands; they are contrasted to the smooth, darker regions known as maria, formed from the ancient lava flows. The lunar highlands are the oldest exposed areas of the Moon, where craters have impacted on older craters, and much of the surface has been obliterated and pulverized from the catastrophic bombardment that took place during the first billion years of lunar history. This period of bombardment was at least 30 times heavier than all the impacts combined during the

Crater Copernicus from far and near, with walls reaching a height of 17,000 feet (5,183 meters). Courtesy Palomar Observatory, California Institute of Technology, and NASA.

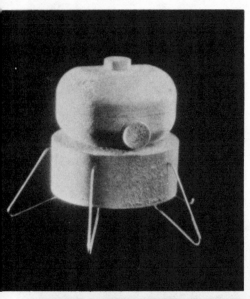

From a balsa-wood model of the Lunar Excursion Module to the real one, Eagle, of Apollo 11. The astronaut is Buzz Aldrin. Courtesy NASA.

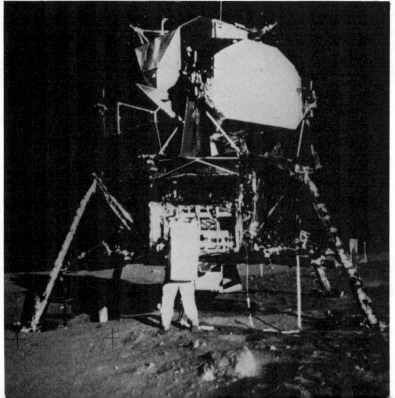

next 3.6 billion years of lunar history—up to the present. The lunar highlands have offered up the oldest rocks to the Apollo astronauts, which is amazing in itself, considering the fact that over time so much of the old rock was churned and buried in 328 feet (100 meters) of soil and in a layer of rubble (breccia) up to 15.5 miles (25 kilometers) thick—over twice the depth of the deepest ocean on the Earth.

☆ ☆ ☆ ☆ ☆

APOLLO KNOWLEDGE

Long-standing Footprints

Footprints left on the Moon by Apollo astronauts will remain visible for at least 10 million years, plenty of time for humankind to visit a nearby star with a suitable planet and leave footprints on its surface.

Life on the Moon

The Apollo 12 astronauts—Conrad, Gordon, and Bean— brought back life from the Moon: a terrestrial bacterium (alpha hemolytic *Streptococcus mitis*) found on a piece of foam inside the television camera of the Surveyor 3 lander. This bacterium survived for 2½ years on the lunar surface and was probably subjected to temperatures inside the camera of up to 70 degrees Celsius (158 degrees Fahrenheit) as well as to the extreme vacuum and dry conditions. *Streptococcus mitis* is a benign inhabitant of the human respiratory tract, and many are dispensed into the air during normal talking. This hearty life brought back from the Moon may all have started with an Earthbound comment or sneeze.

The Oldest Moon Rock

A fractured rock found on the Moon was dated at 4.6 billion years, making it the oldest found during the Apollo missions. Its age dates back to the time the Moon and Earth were formed. Now it is the oldest rock on Earth. This rock was 1 billion years old before the first single-celled microorganisms lived on Earth.

The Youngest Moon Rock

The dark Moon maria (''seas'') formed by the great lava floods over 3 billion years ago contain the Moon's youngest rock. The youngest Moon rock returned to Earth was found in the Hadley Rille area by the Apollo 15 crew and was dated at 3.1 billion years—just about when the earliest bacterial life began on Earth.

The "Great Scott" Rock

The second largest moon rock brought back to Earth was the 21-pound (9.5-kilogram) ''Great Scott'' rock that Apollo 15 astronaut David R. Scott found near the 1,000-foot (305-meters)-deep Hadley Rille. Dated as 3.3 billion years old, it is a volcanic basalt probably formed when the Imbrium Basin was flooded with lava. This rock has a micrometeorite crater in its center, which demonstrates how the process of lunar erosion takes place over millions of years.

Mining the Moon

After millennia of wonder and speculation, we know what the Moon is made of—this is perhaps the greatest practi-

A terrestrial bacterium survived on the Moon inside the television camera of the Surveyor 3. The camera was brought back by Apollo 12. Courtesy NASA and the Lunar and Planetary Institute.

The oldest Moon rock, 4.6 billion years, is now the oldest known rock on Earth. Courtesy NASA.

Big Muley, the largest rock brought back from the Moon, weighed almost 26 pounds (12 thousand grams). Courtesy NASA.

Astronaut Harrison (Jack) Schmitt of Apollo 17 is working alongside the Rover. He first saw the famous orange soil at this spot. Courtesy NASA.

cal legacy of the Apollo program. Lunar rock and soil contain plagioclase, anorthosite, ilmenite, and silicon, to name a few minerals, and these can be processed to give aluminum, titanium, iron, magnesium, silica, and oxygen. It is simply a matter of economics and time as to when lunar resources will be utilized, but the condition of nonrenewable resources on the Earth—currently demonstrated by the world oil supply—may well make a permanent Moon base and Moon mining a reality sooner than most current predictions. If Solar Power Satellites are built within the next century, most of the materials to build them will no doubt come from the Moon, where low gravity and new electromagnetic launching techniques could economically transport large amounts of material into space for refinement and construction. If 1 million tons of typical lunar soil (the so-called "fines") were processed, they would yield approximately 56,000 tons of aluminum, 139,000 tons of iron, 56,000 tons of titanium, 44,000 tons of magnesium, 194,000 tons of silicon, and 139,000 tons of oxygen. The whole scenario of space industrialization and colonization that has received so much press in the last several years depends on lunar and asteroid resources. The Moon's resources are close, and now that we know what they are, we will be using them. Think of a Moon Ford factory turning out hundreds of Lunar Rovers.

Moonquivers

Moonquakes are so weak in comparison to earthquakes that "moonquivers" might describe them more aptly. About 3,000 moonquakes occur each year, while about 800,000 earthquakes of the same intensity occur for the same period. Most moonquakes are associated with tidal forces caused by the Earth's gravity as it increases during the Moon's orbit around the Earth. There is another kind of moonquake, however, that occurs in groups and seems unrelated to Earth-Moon orbital distance. It is speculated that this kind of moonquake is related to the shifting of molten or partly molten rock in the Moon's core. Moonquakes occur much deeper in the Moon's interior than do earthquakes on Earth—about 370 to 500 miles (600 to 800 kilometers), and their reverberations last much longer. The first artificial moonquake that NASA created by crashing the Apollo 12 Lunar Module onto the Moon (November 1969) set the seismometer vibrating for more than two hours. The cracked and broken rocky outer layer of the Moon caused the signals to bounce around for hours. One project scientist said that the Moon rang like a bell. This very active and surprising seismic activity, however, does not provide a good indication of moonquake energy. The average moonquake does not release much more energy than a firecracker, and all the moonquakes in a year would not produce more energy than a modest Fourth of July fireworks display.

Quake or Strike?

Moon-based seismometers pick up meteorite strikes as well as moonquakes, but they produce slightly different readings so that the trained eye can tell them apart. Because the interior of the Moon is so quiet compared to that of Earth, even the impacts of small meteorites can be detected. Any meteorite weighing more than 22 pounds (10 kilograms), about the size of a big grapefruit, would be detected no matter where it fell on the Moon. About 100 meteorite impacts are registered each year, but unlike the moonquakes, their combined energy would pack quite a wallop.

The Daily Strike

The Apollo 12 seismometer set up on the Ocean of Storms was designed and expected to operate for one or two years. It exceeded all expectations by operating continuously from the time Charles Conrad and Alan Bean set up the equipment in November 1969 until September 1977, when it was shut down for budgetary reasons. During that time it registered almost 2,300 moonquakes and meteorite impacts—averaging out to almost 1 strike a day.

The Moon's Red Glow

A strange red glow has often been observed on the Moon's surface when it comes closest to the Earth (perigee). These observations have been made for centuries, and the cause has always been a mystery. The seismic equipment left by Apollo 12, however, may have solved the mystery. Seismic readings indicate that moonquakes occur at the same time the red glow appears—and both when tidal forces are strongest between the Earth and the Moon. These tidal forces may cause the lunar surface to "pop," which in turn causes moonquakes that release trapped gases beneath the surface. The ancient orange-red glow observed for centuries may be the escape of these gases—a kind of lunar burp.

☆   ☆   ☆   ☆   ☆

FUTURE OF THE MOON

The "Smaller" Moon

The last of the dinosaurs saw the basic Moon that we see today. The only differences, unseen from Earth, are some man-made hardware and human footprints on the surface. Future life on Earth may see a "smaller" Moon, since it is receding from Earth about 1 mile (1.6 kilometers) every 28,000 years owing to the results of tidal bulges and gravitational forces. If this rate continues, the Moon from Earth will appear 15 percent smaller in about 1 billion years.

Full-Moonless February

The month of February will be very unusual 2.5 million years from now. There will be the rare event of no full Moon in February in that far-distant year. Full-Moonless February occurred in recent history, however—in 1866, the same year that the principles of heredity were stated by botanist Gregor Mendel.

No Man-made Craters

The first lunar base will probably be built to start up and support Moon mining, and the Moon-mining base will be the foundation for economical space industrialization and colonization, which will build space manufacturing facilities and town-sized habitats at certain locations between Earth and the Moon.

According to some scientists, the initial purpose of the Moon-mining and resource-launching base will be to sustain a space population whose task will be to build the Manhattan-sized Solar Power Satellites that will collect the Sun's undiluted light and microwave it down to the Earth, where it will be converted into electricity for terrestrial use. The Moon base, with perhaps 100 people initially, would mine and launch about 1 million tons of lunar rock and soil each year, which—with the help of the free and abundant solar energy in space—would be transported to the space refining facility and made into metal structural members and other products. Visions of man-made craters left behind by lunar mining operations are unrealistic. The amount of lunar material needed to build a spherical-type space habitat large enough for 10,000 people has been estimated at about 3.6 million tons. The excavation site for this amount of material would have a depth of about 13 feet (4 meters) and be a square area, each side of which is 2,460 feet (750 meters)—more like skimming than digging, so no man-made craters will be left behind.

Moon-Base Housing

The living and working quarters of the first Moon base will be made from the fuel and payload sections of the rocket, and part of the first payload will be the lunar soil-moving and soil-blowing equipment. One of the first tasks in setting up the Moon base will be to bury these rocket sections under several feet of lunar soil by using a soil-blowing vehicle that works on the same principle as a snow blower. The thick covering of lunar soil over these Moon houses will protect the Moon workers against lethal solar flares and penetrating cosmic rays. The Moon soil will also insulate and regulate the temperature of the living space. Apollo missions found temperature differences of 260 degrees Celsius (500 degrees Fahrenheit) between lunar night and lunar noon. The temperatures near the equator were 110 degrees Celsius (230 degrees Fahrenheit) at lunar noon and −179 degrees Celsius (−290 degrees Fahrenheit) just before dawn of the two-week day. Apollo 15 learned that just 3 feet below the surface there was only a temperature variation of a few degrees, which proves that a Moon house insulated with lunar soil will be much more energy-efficient than one insulated with fiberglass—cheaper, too.

Moon Babies

Moonsteaders will have to return to Earth every few years—either that or stay on the Moon for the rest of their lives. The reason: Their bodies will lose muscle mass because of the light gravity, and their bones and metabolisms will change, so that after several years in the low lunar gravity, they may not be able to return to Earth safely. Moon-born babies will also be affected by the low gravity. Indeed, their bodies may well become taller and more slender because of their embryonic development in a low-gravity womb. This should please expectant mothers and life insurance companies.

Lunar Slowdown

The Moon takes an additional two thousandths of a second each year to circle the Earth—about the same time that it takes a balloon to pop. Lovers of the far future will have less moonlight time to share—and it will be dimmer, too, as the Moon gradually pulls farther away from the Earth.

☆　☆　☆　☆　☆

LUNAR INFLUENCES

The Oyster Connection

Evidence is accumulating that points to a direct connection between the lunar cycles and human behavior. This should come as no great surprise to anyone who knows that, like the surface of the Earth, the human body contains about 80 percent water and 20 percent solid matter. Why should there not be "biological tides" just as there are ocean, atmospheric, and land-mass tides? Some researchers, doctors, and scientists believe that, indeed, there are high and low tides in all life, and that the high tides occur at new and full Moon. This is when the Moon affects animal and human behavior most. One interesting experiment involved some East Coast oysters that were air-freighted to the Midwest and put into a controlled environment of seawater, completely cut off from any tidal clues, including the Sun. These oysters had opened their shells at high tide on the East Coast, but soon gave that timing up in the Midwest. Instead, their shells opened when the Moon was at zenith overhead in their new location, which would have been the new high-tide time if they were not land-locked. This was proof that the Moon's gravitational pull directly affected the oysters' behavior. If you have not already noticed, oysters and people have much in common.

Moon Madness

Recent research indicates that there is a definite relationship between human aggression and the lunar month (from new Moon to new Moon; about 29.5 days). Increases in rape, robbery, assault, burglary, and disorderly behavior occurred during full Moon. Also, patient admissions in state mental hospitals and emergency rooms increased. There is growing evidence that the lunacy myth has a basis in fact and that people, to varying degrees, are affected by tidal stresses of the Moon.

Body Tides

When the Moon is full or new, the pulse rate increases for people who are under stress. Blood studies show that chemicals in the blood responsible for increasing pulse rate rise sharply for this group just after full and new Moon. These cardio-accelerating chemicals are not found, however, in people in unstressful situations just after full and new Moon, but they are at other times. Another finding: People bleed more at certain times in the lunar cycle. Increased bleeding during surgery—for example, tonsil and adenoid operations—occurs around full and new Moon. Also, the frequency of bleeding-ulcer attacks goes up at a significant rate. Why? One theory is that a change in the Earth's electromagnetism, brought on by the Moon's rotation around the Earth, is somehow picked up by our nervous systems, and our bodies respond in many ways.

Sex and the Moon

There may be more than a romantic connection between the Moon, love, and sex. The connection is certain with regard to sea urchins, whose reproductive cycle follows the lunar cycle exactly. On the average, the menstrual cycle for women is exactly the same duration as the lunar month—29.5 days; and the human gestation period is exactly nine times the lunar month—265.8 days. Studies have shown that there are more births during full and new Moon, when the gravitational influence is the strongest, and that more male children are born after full Moon and more female children born after new Moon. Slowly, much Moon lore (for example, the Navajo belief that there are more births at full Moon because of the Moon's pull on the embryonic water) and superstition is being proven statistically. What does all this mean? It means that the Moon has for eons governed the movements of the ocean tides, tides that were part of the origin and evolution of life. It is therefore not surprising that the human reproductive system follows the lunar cycle—an astronomical clock rather than a biological one—and that lovers walk in the moonlight.

METEORS, ASTEROIDS, AND COMETS: THE CELESTIAL SIDESHOW

☆　☆　☆　☆　☆

METEORS AND METEORITES

A Famous Meteor Shower

Meteor showers occur when the Earth in its orbit passes through a meteoroid swarm or stream in another orbit; there are at least a dozen major such encounters and showers each year. It has recently been determined that meteor streams have the same orbits as comets, and that most meteors are bits of debris left behind by comets. The most spectacular display of a meteor shower in recent history occurred on November 17, 1966—the annual Leonids. Visible meteors flashed at over 2,000 per minute during the shower's peak. At this rate, an observer would miss seeing over 30 with each blink.

Space Dust

Most of the 100 million meteors that enter the Earth's atmosphere each day burn up and filter down to Earth as dust. The total weight of this dust to fall in one year is estimated to be 4 million tons—enough to cover the Earth with a layer 1 inch thick in 5,000 years. When you next take a dustcloth to the bookshelves, remember that some of it is ancient star dust, perhaps as old as the solar system itself—4.6 billion years.

The Tunguska Blast, Siberia

The most dramatic and awe-inspiring meteorlike event in recorded human history was the great Tunguska fireball and explosion June 30, 1908, which took place in an isolated region of Siberia 40 miles (65 kilometers) north of Vanavara. A great fireball, brighter than the Sun, was witnessed in the region; then it exploded about 5 miles (8 kilometers) above the surface with a blinding terminal flash. The power and destruction of the blast was incomprehensible. Its power has been estimated at 30 megatons, 1,500 times as great as Hiroshima, and comparable only to the heaviest hydrogen bomb. The violent blast was registered in Moscow, Germany, England, and even Washington, D.C., thousands of miles away. Five hundred miles (800 kilometers) away the thunderous explosion was heard. People up to 40 miles (65 kilometers) away had to shield their faces from a fierce heat wave, and their clothing was singed. Windows were blown out up to 50 miles (80 kilometers) away, and people were knocked off their feet up to 100 miles (160 kilometers). Close to the blast, whole herds of reindeer were killed, instantly burned to death. Four hundred miles (640 kilometers) away, horses stumbled and fell to the ground. Trees up to 30 miles (48 kilometers) from the blast were blown down, and they were scorched up to 14 miles (22

The great Leonid meteor shower of 1966, when about 2 thousand meteors flashed across the sky each minute. Many were faint streaks, but others were fireballs as bright as the Moon. Courtesy C.F. Capen, Braeside Observatory.

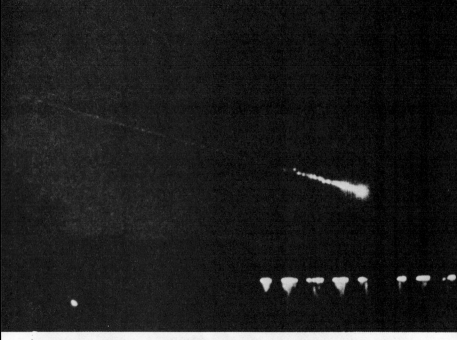

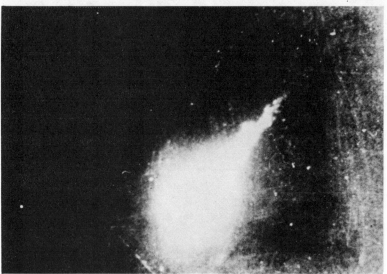

Two bright fireballs captured on film: (Top) April 1966 fireball over Springfield, MA; and (Bottom) the Boveedy fireball exploding over Northern Ireland, 1969. Courtesy Smithsonian Astrophysical Observatory: and the Armagh Planetarium.

kilometers). The total destroyed area of Tunguska amounted to 1,200 square miles (3,100 square kilometers), more than the combined areas of New York City, Philadelphia, Los Angeles, and San Francisco. Assuming that this event was caused by a meteor, and there is still considerable debate about this, its weight has been estimated at 1 million tons, its diameter at about 300 feet (90 meters).

After the event there was a "black rain," caused by a huge cloud of dirt and debris thrown up into the atmosphere by the blast to a height of 68,000 feet (20,000 meters). The smaller particles of this cloud rose even higher into the stratosphere and caused unusual high-time lighting effects around the world. The Sun lit up this dust over the North Pole, and newspapers could be read at midnight in Europe and Western Siberia.

Many scientists and researchers believe that it was the head of a comet striking the Earth; others attributed it to antimatter. Whatever the cause, it is the one event in recent history that gives humankind a hint of the inhuman power of cosmic forces.

The Arizona Meteor Crater

The best-known, best-preserved impact crater on the Earth is the Arizona Meteor Crater, also referred to as Barringer Crater and Coon Butte. About 22,000 years ago, a mammoth iron meteoroid, weighing close to 100,000 tons and having a diameter of perhaps 100 feet (30 meters), broke through the atmosphere and struck the Arizona desert at over 25,000 miles (40,000 kilometers) per hour with an impact force of 2 million tons of TNT. The great blast displaced millions of tons of rock and dug a crater bowl over 4,000 feet (1,219 meters) in diameter and almost 600 feet (183 meters) deep, pushing the walls 150 feet (46 meters) above the surrounding plain. But such a cataclysmic force is dwarfed when the Moon's Copernicus crater is contemplated. Copernicus Crater is 56 miles (90 kilometers) in diameter, almost 75 times larger than Meteor Crater, and the meteorite that blasted it was at least a few kilometers in diameter. The blast would have been clearly visible from Earth had there been eyes to see it. While not as bright or dramatic, the Arizona blast would have been visible from the Moon had there been eyes there to see it—a visible flash from the impact spot, lasting several seconds.

Meteor Brightness

A meteor as bright as Sirius, the brightest star in the sky, would have a diameter of one fiftieth of an inch or less and weigh less than one hundredth of an ounce. A meteor the size of a walnut would appear as bright as a full Moon. The Sumava meteor, which appeared over Czechoslovakia in December 1974 as a brilliant fireball, was a 200-ton boulder before it was destroyed by a series of midair atmospheric explosions. Each explosive flash was 500 times brighter than the full Moon. The great Tunguska fireball of 1908 had a dazzling fireball brighter than the Sun.

The Earth's Star Wounds

Astroblemes, literally "star wounds," are another name for impact craters on Earth. If it were not for the atmosphere's protective shield, as well as erosion, mountain building, and other forces at work on Earth, the surface would appear crater-strewn like the Moon or Mercury or Jupiter's satellite, Callisto. In fact, many large craters on Earth probably remain undiscovered in the jungles of South America, Africa, and

The Arizona Meteor Crater compared in size to Lower Manhattan.
Courtesy Smithsonian Astrophysical Observatory.

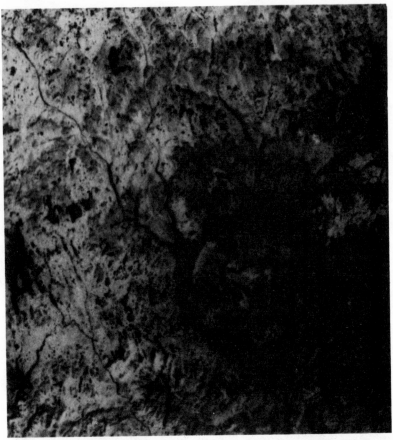

One of the largest astroblemes (ancient meteor craters) on Earth—off the eastern shore of Hudson Bay. Courtesy Smithsonian Astrophysical Observatory.

Asia and in the unexplored deserts of the world. Some crater remnants may never be recognized, especially in mountainous regions. However, craters continue to be found. The Wolf Creek Crater, almost as large as Arizona's Meteor Crater, was discovered in 1947 in northwestern Australia. A Skylab photo shows a highly eroded crater in Canada—a circular lake is all that remains of this 210-million-year-old impact crater that once was over 40 miles (65 kilometers) in diameter. The largest suspected astrobleme on Earth is off the eastern shore of Hudson Bay. Scientists believe that the circular arc formed by the Nastapoka Islands there may be part of an ancient meteor crater rim left by the impact of a giant asteroid in prehistoric times. But even this crater, measuring 275 miles (442 kilometers) in diameter, is just a baby when compared to the Caloris Crater Basin on the planet Mercury, which is more than 745 miles (1,200 kilometers) wide.

Raining Solid

An estimated 100 million meteoroids enter the Earth's atmosphere each day, and this does not include the uncounted billions of micrometeorites that drift down undetected. From this tremendous number, only 500 fall to the Earth each year, and only 10 of these are recovered, since most fall and sink into the oceans.

The largest meteorite in the world, at Hoba West, near Grootfontein, South-West Africa, remains where it fell, being much too large to be moved. It measures 9 by 8 feet (2.7 by 2.4 meters) and weighs 132,000 pounds (almost 60 metric tons), as much as nine adult bull African bush elephants. The smallest micrometeorites, by comparison, collected by rockets at high altitudes, are less than 1 micron in diameter (a micron is one thousandth of a millimeter!).

A Cosmic Long Shot

No person has been killed in modern times by a meteorite fall—at least not in the authenticated records. In 1511, however, folklore has it that a monk was killed by a meteorite in Cremona, Italy. Animals have been less fortunate. An 1860 fall in Ohio killed a calf, and in 1911 a fall in Egypt killed a dog.

The only human injury on record as caused by a meteorite took place in Sylacauga, Alabama, November 30, 1954. Mrs. E. Hulitt Hodges was napping on her living-room couch when a 3-pound meteorite crashed onto the roof of her home, tore through the living-room ceiling, bounced off a console radio, and struck Mrs. Hodges on the left thigh, severely bruising her. The odds against Mrs. Hodges' being hit by this celestial stone were about 100 billion to 1. The odds against you or me being hit in any given year are 10 trillion to 1.

☆   ☆   ☆   ☆   ☆

ASTEROIDS

Trimming Off Pounds

A future astronaut who works in the mining operations of the Asteroid Resource Retrieval Corporation weighs 150 pounds on Earth. When on duty, this astronaut is the envy of all weight-reducers on Earth. On the asteroid Eros, he weighs 4 ounces, and on some of the smallest asteroids in between Mars and Jupiter, he weighs about 1 ounce. Like all weight reduction, however, it is temporary.

The First Asteroid

Ceres is the largest of the asteroids and was the first to be discovered—on the first night of 1801—by an Italian astronomer named Giuseppe Piazzi. The discovery was accidental; Piazzi saw it while compiling a star catalog and realized it was an intruder. This asteroid is an orbiting island of jagged rock, about 620 miles (1,000 kilometers) in diameter, and its sphere could fit into the state of Texas, with just some minor shifting of borders. Ceres has a surface area of about 700,000 square miles—more than the areas of Alaska and Nevada combined.

Odd-Shaped Eros

The asteroid Eros, a cigar-shaped rock 15 miles (24 kilometers) long and 5 miles (8 kilometers) wide, has almost the same dimensions as Manhattan. Its oblong shape was observed in 1931, when it came to within 16 million miles (26 million kilometers) of Earth, and it was closely studied in 1975 when it passed in front of a star. Eros, only 4.4 miles (7 kilometers) thick, tumbles end over end as it circles the Sun every 21 months. To two lovers walking hand in hand on the surface of Eros, there would be no horizon, and they could lean over either of two long edges and view the starry sky below.

Pop Fly

A baseball hit on the surface of Eros would immediately reach escape velocity, fly off the asteroid, and go into orbit around the Sun—a pop fly that would never come down.

Asteroid Obstacle Course

The nineteenth century was the century of the asteroids. The first was discovered on January 1, 1801, and more than 300 had been discovered by 1890, at which time photographic time exposures were used in the search. Over 2,280 asteroids have been assigned permanent numbers, and about another 1,000 have temporary designations. It is estimated that perhaps 100,000 observable ones remain uncharted, so the majority of these bodies may be awaiting discovery. One statistic, which considers inclusion of smaller and smaller bodies, predicts that there should be 22 million—even thousands of millions—when the smallest fragments are put in the count. It would seem virtually impossible for future astronauts to keep track of all of them while their ships navigate the asteroid belt. But after all, Captain Video and Han Solo did it.

Closest Approach

The small asteroid Hermes, with a diameter of about 2,000 feet (610 meters), holds the record for the closest approach to Earth, passing within 485,000 miles (780,000 kilometers) in 1937—only twice the distance to the Moon. At some future date, Hermes could come even closer and pass between the Moon and the Earth. Other asteroids—for example, Amor, Icarus, Apollo, and Adonis—also have orbits that bring them relatively close to the Earth-Moon system. Their proximity underscores the long-term practicality of someday mining them for their ore deposits. Several studies indicate that billions of dollars' worth of ore could be obtainable from asteroids, enough to supply terrestrial iron needs for decades.

The Red-Hot Asteroid

The asteroid Icarus has the smallest orbit known for a minor planet, the period of which is about 1.3 years. It comes within 17 million miles (27 million kilometers) of the Sun, about half again as close as the planet Mercury at its closest, and goes out beyond the orbit of Mars, which keeps it inside the main asteroid belt. Icarus has a day that is 2.3 hours long and moves about its orbit at speeds exceeding 50 miles (80 kilometers) a second. This rock passed within about 4 million miles (6.4 million kilometers) of Earth in 1968 and was observed to be ball-shaped, about 1 kilometer across. It passes near the Earth every 28 years or so, but, since it was discovered only in 1949, we have had just the one opportunity to observe it up close so far. When Icarus is closest to the Sun, its surface temperature climbs to about 400 degrees Celsius (752 degrees Fahrenheit), which makes its sunward surface red hot—an orbiting oven.

Bright Vesta

Vesta, the fourth asteroid to be discovered (after Ceres, Pallas, and Juno) in 1807 by Heinrich Wilhelm Olbers, a German doctor who was also an amateur astronomer, is the brightest of all the asteroids. This is true not just because of its relatively large size, in the 248–310-mile (400–500-kilometer) diameter range, but also because of the high reflective quality of its surface, probably composed of basaltic minerals. At times Vesta can be seen with the unaided eye as a faint speck of light. The blackest asteroid known is 324 Bamberga, nearly as dark as soot from a fireplace.

COMETS

Celestial Fossils

Comets have been called "celestial fossils"—survivors from the formation of the solar system 4.6 billion years ago. It is believed that these millions of small bodies were left in the outer solar system, beyond primordial Neptune, where gravity was weak, when the sunlight's pressure drove off the gassy remains of the primordial cloud from which the Sun and planets condensed. Comets, with their ice-filled structures and eccentric orbits, show what condensations of the early solar system were probably like. A sun-grazing comet, with its nucleus, coma halo, and tail aglow, gives an idea of what the protoplanets may have looked like when the just-born Sun vaporized their gassy outer envelopes. The Earth may have looked like a giant comet, its tail streaming outward, away from the Sun's youthful light.

Unlikely, but . . .

A huge comet could strike the Earth, although the probability is low. Given enough time, however, the probability increases. If there ever were a head-on collision between the Earth and a large comet, the relative speed would be about 80 kilometers a second (178,848 miles per hour). This tremendous speed and the resulting cataclysmic impact explosion would lift a large volume of the Earth's atmosphere into space, with disastrous consequences for life near the impact area.

A Cosmic Ticking

Halley's comet has been a cosmic clock for the Earth for at least 2,500 years—and probably much longer—appearing about every 76 years. Before its appearance in 1758, it was referred to as "the great comet" or "the bright comet" in many cultures and languages over the centuries. But in that year it was named Halley's comet to honor posthumously the English astronomer who first predicted its return. Edmund

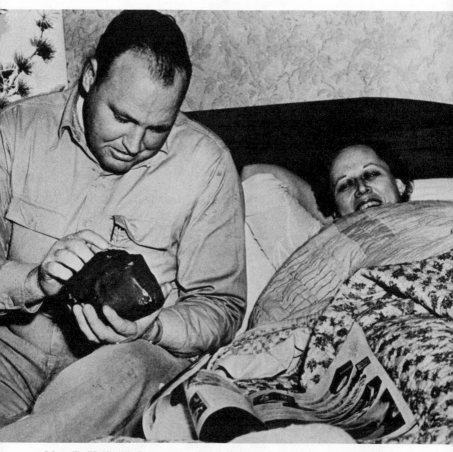

Mrs. E. Hulitt Hodges recuperating from her meteorite injury. Her husband holds the cosmic stone that struck her. Courtesy The Birmingham News Company.

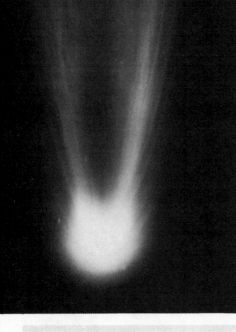

Halley's Comet has ticked off its 76-year orbital period for more than 2,500 years, and it is still going strong. It was depicted in the Nuremberg Chronicle in 684. Much was learned about Halley's Comet during its 1910 passage, at which time its tail became 100 million miles (161 million kilometers) long. Courtesy Mount Wilson and Las Campanas Observatories, Carnegie Institution of Washington, and Yerkes Observatory.

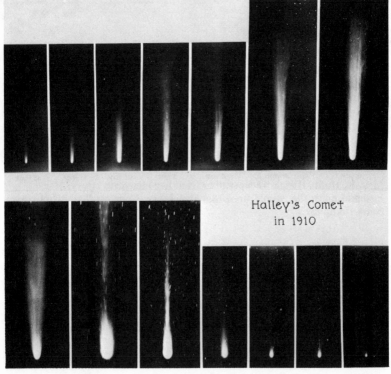

Halley's Comet in 1910

Halley worked out the orbit of the bright comet of 1682 and realized that it was very similar to the orbits of the 1531 and 1607 comets. His calculations, along with those of Sir Isaac Newton, led him to predict that it would again visit the Sun and Earth in about 1758. On Christmas Day, 1758, seventeen years after Halley's death, the comet was rediscovered by an amateur astronomer in Germany, and astronomical knowledge took a giant leap forward.

Halley's comet has since been traced back, through Chinese and Japanese written records, to 467 B.C., the year that Athens became the leading member of the Delian League. It has thus made 31 recorded swings around the Sun. The date of its original appearance is not known and never will be. From the study of written records, this great comet has shown no signs of wear between 467 B.C. and A.D. 1910, so it has probably had an extremely long life and innumerable visits to the inner solar system.

The glow of its coma halo was visible from the Earth in 240 B.C., when Roman literature began with a translation of a Greek play into Latin. Its tail probably swept across the sky in 2000 B.C., when the signs of the zodiac were established in Mesopotamia. It blazed near the Sun a few years before the birth of Christ (12 B.C.), about the same time that astronomy rapidly developed in China under the Han Dynasty. Three passes later, in about A.D. 218, it marked the end of the Later Han Dynasty in China. In about A.D. 370, it marked the beginning of the Third War between Rome and Persia, and it appeared in the sky again to mark the creation of Anicius Boethius's work, *The Consolation of Philosophy,* in about A.D. 530. Halley's comet is depicted in the Nuremburg Chronicle in 684, and it alarmed the Saxons in 1066, as William of Normandy invaded England. Centuries later, in 1301, its passing marked Edward II's subjugation of Scotland. No other comet has been seen by so many people over such an expanse of time, as it visually ticked off the centuries with a sweep or two of its 100-million-mile-long tail.

Much more attention was paid to this great comet after Halley predicted its return in 1759. Mark Twain (Samuel Clemens) was born under it in 1835 and died under it in 1910. During this 1910 appearance, much was learned about the comet for the first time. The comet's nucleus was estimated to be between 10 and 30 miles (16 and 48 kilometers) in diameter and came as close as 15 million miles (24 million kilometers) to Earth. The orbit was more accurately defined. It came as close as 56 million miles (90 million kilometers) to the Sun during its inner-solar-system swing and traveled out beyond the orbit of Neptune to a distance of about 3.3 billion miles (5.3 billion kilometers) at its farthest point in orbit, where it was in 1948. About half its orbital time (38 years) is spent on that part of its orbit which lies beyond Neptune—a small percentage of the entire orbit.

As Halley's comet approached the Sun in 1910, its tail grew at the rate of a half million miles per day and was estimated to be 100 million miles (161 million kilometers) at one time. During this passage, the comet passed directly between the Earth and the Sun, and some people predicted that its silhouette would be visible as it passed in front of the Sun. Nothing was seen, however. The Earth passed through the rarefied tail of Halley's comet on May 21, 1910. Because of sensational newspaper stories at the time, there was some minor mass hysteria. People heard that there were some poisonous gases in the tail and feared that the world might come to an end. But nothing happened. The Moon was nearly full at this time, so any unusual atmospheric effects were washed out by the bright moonlight.

Halley's comet is expected again in the early spring of 1986. Studies suggest that it will not be a brilliant sight from the Earth, or at least most regions, and binoculars may be needed. But comets are unpredictable. Remember Kohoutek?

NASA and the European Space Agency hope for a close-up view, however, if an unmanned spacecraft is launched on schedule. It would fly by Halley's comet in 1985 and shoot a probe into the comet's head, where a camera could photograph an object as small as a baseball in a comet's nucleus. Humankind may thus get its first close-up of the world's most famous comet.

Comet Life

Since the famous laboratory experiments by Dr. Stanley Miller in 1952, when he produced amino acids, the basic building blocks of life, by passing an electric current through a mix of gases, other scientists have identified organic molecules in comets and meteorites. Since comets have the basic material for life, there has been speculation that comets falling to Earth in early history provided the essential elements for the origin of life. It has also been suggested that comets themselves are a suitable environment for life. Astronomer Fred Hoyle has further speculated that the equivalent of viruses are formed on comets and in interstellar clouds and that new viruses may come to earth through the atmosphere by passing comets, thus causing pandemics. Other scientists, however, are highly skeptical about this.

What Price Beauty?

Deep in space, away from the Sun, a comet is a cold, frozen chunk of gases and solid particles. But a comet dons its celestial raiment and becomes beautiful when it is close to the Sun—when the Sun's radiation heats up and vaporizes the gases, which then begin to glow in the comet head's nucleus and coma halo. The Sun's light exerts a force on the glowing gas particles, and some of them are blown away from the head to create a tail. Each time a comet swings around the Sun, then, it loses a small portion of itself through this process. Comets burn out a bit more each time they offer us their celestial display. Assuming that Halley's comet loses just a millionth of its mass on each pass around the Sun and that it has existed for the entire age of the solar system, about 4.6 billion years, its original size would have been larger than the Sun!

The Lucky Discovery

Charles D. Perrine, as astronomer at Lick Observatory, had a lucky year in 1896—the same year that Henry Ford made his first automobile. He had discovered a comet earlier in the year and requested up-to-date information on its position from a fellow astronomer, so that he could observe it once more. The words in the telegram were somewhat garbled, however, and the position given was wrong by more than 2 degrees. Perrine, unaware of the error, pointed his telescope to the coordinates given in the telegram. Behold! By an amazing coincidence, he viewed an entirely new comet right in the center of the eyepiece!

A Time So Rare

When a comet approaches the Sun, it grows to a tremendous size without gaining mass. What was a frozen gas ball of debris in the outer reaches of the solar system becomes a glowing wonder complete with a nucleus, a coma halo, and, usually, an impressive tail. Comets owe their visible splendor to our Sun, whose radiation brings them to life after their frozen hibernation. If the hydrogen envelope of a comet, which can only be observed from outside the Earth's atmosphere, and the tail are included in the measurement, many comets occupy a space several times larger than the Sun. But all the estimated 100 billion comets put together would weigh no more than the Earth. The density of a typical comet is far below that of water or air—one estimate puts it at a quarter of a millionth the density of air or less. In other words, a comet's beauty is more rarefied than any vacuum produced in our laboratories on Earth. The tenuous nature of comets is apparent when one observes stars shining undimmed through comet heads. Such a nighttime view is a rare event in itself, since the observer would be seeing ancient starlight shining through newborn comet-reflected sunlight, the former perhaps thousands of years old, the latter just minutes old—a rare, beautiful mix of visible time.

Cloud of Comets

There are about 100 billion comets that circle the Sun and planets in a halolike cloud known as Oort's cloud from about 12 billion to 10 trillion miles (19 billion to 16 trillion kilometers) out from our solar system. Occasionally some comets in Oort's cloud are influenced by the gravity of a planet or a nearby star and leave their faraway orbits to venture into the solar system and make all earthlings feel humble. If we could locate, identify, and name all of the 100 billion comets, there would be 24 personal comets for each woman, man, and child on Earth.

The Comet Head Hunter

Comets that come to visit our Sun from the cometary cloud run the risk of capture (or expulsion) by the giant planet of our solar system, Jupiter. Time and time again Jupiter has proven its strength and dominance by changing comet paths or actually capturing them. The giant planet is believed to have a captured entourage of about 30 short-period comets. Comet Brooks was captured in 1886 when it passed within about 55,000 miles (88,000 kilometers) of the planet's surface; its orbital period was changed from 27 years to 6.8 years. This may be a case of where a prospect was lured into the system, with several successive encounters reducing the orbit by stages over a long period of time.

Comets on the Near Curve

All comets accelerate as they fall toward the Sun and then whip around it at fantastic speeds; the closer they come to the Sun, the faster they go. The great comet of 1882 passed within 300,000 miles (483,000 kilometers) of the Sun's surface and reached a speed of 900,000 miles (1,448,000 kilometers) per hour. In 1965, Comet Ikeya-Seki swung as close as 290,000 miles (466,000 kilometers) of the Sun and accelerated to over 1 million miles (1.6 million kilometers) per hour. Undoubtedly, some comets of the past have not been able to take the curve and have flown into the Sun—at an estimated speed of 1.4 million miles (2.25 million kilometers) per hour. The Sun-grazing comets that make it, however, pay the price for their daring spree—they are broken up by strong tidal forces of the Sun. The nucleus of the great comet of 1882 divided into four parts after its close encounter and proceeded single file as a comet group. Comet Ikeya-Seki split in two; in fact, some observers claimed that it split into three parts. That some comets reach a speed of over 1 million miles (1.6 million kilometers) per hour should impress even jet pilots. And all of us are assured that it is a respectable cosmic speed when we learn that the Earth travels a mere 67,000 miles (108,000 kilometers) per

Comet Ikeya-Seki swung to within 290 thousand miles of the Sun in 1965—
a Sun-grazer. It accelerated to more than 1 million miles (1.6 million
kilometers) per hour. Lick Observatory photo.

The Great Comet of 1843, with its famous tail that stretched half-way across the sky. It was so long (about 500 million miles; 800 million kilometers) that it could have gone around the Earth 20 thousand times. Drawn by Charles Piazzi Smyth at the Cape Observatory, March 1843. Courtesy Durban Public Library, South Africa, and Brian Warner.

hour in its orbit around the Sun, and that the fastest a human has traveled in space is 24,791 miles (39,889 kilometers) per hour, when the Apollo 10 spacecraft reentered the Earth's atmosphere.

The Longest Tail

The great comet of 1843 was deserving of its greatness: Its tail stretched halfway across the Earth's sky, and astronomers estimated it to be about 500 million miles (800 million kilometers) long—just over Jupiter's distance from the Sun. If this cosmic tail were wrapped around the Earth, there would be enough of it to circle the equator 20,000 times.

Lucky Comets

Comets were regarded with superstitious awe and were thought to signify impending disaster before the eighteenth century, when Edmund Halley discovered that they revolve around the Sun in large orbits. Pope Calixtus III asked everyone to pray for deliverance in 1456—"from the devil, the Turk, and the comet." The devil, of course, was always at large; the Turks were threatening to invade Europe; and the comet appearing at the time, later named Halley's, proved to those of the Middle Ages that evil was near. Today most people consider the sight of a comet good luck, and well they should. Dutch comet authority Jan Oort estimates that only 1 in 100,000 can be observed. Even so, the chance of every living person seeing at least one large and bright comet during his or her lifetime is good. So . . . good luck on this chance of a lifetime.

5

THE STARS: TWINKLE, TWINKLE ... BOOM!

☆ ☆ ☆ ☆ ☆

THE STARS FROM EARTH

Counting Stars

We have direct proof of the existence of at least 100 billion billion stars . . . and we're still counting.

The Isolated Stars

Stars, to the human eye, appear to crowd the clear night skies, but they actually are separated by enormous distances. In fact, only about 1 part in 100 million of the volume of space is filled with stars. If our Sun were represented by a basketball in New York City, then on the same scale the solar system would have a diameter of not quite 2 miles (3 kilometers), and the next star would be a basketball about 5,000 miles (8,000 kilometers) away in Honolulu, Hawaii.

Too Far to Drive

Alpha Centauri is the nearest star system to our Sun—4.3 light-years away. It is really a triple star system, and the dimmest companion, Proxima Centauri, is slightly nearer to the Sun than the two brighter stars. An Apollo spacecraft would take 850,000 years to reach the stars of Alpha Centauri if its average speed were the same as a 6-day 500,000-mile round trip to the Moon. It would take a VW Rabbit 52 million years to make the same trip at 55 miles per hour, which is equal to 722,000 average lifetimes.

Starlight, Not Too Bright

How much starlight at any one time falls on Earth? The light from all the stars equals about one fifteenth the light of the full Moon or one six-millionth of the Sun's light. If the sum of all starlight could be concentrated in one object, it would equal a 100-watt bulb seen from a distance of 613 feet (187 meters)—about the length of two football fields.

Stellar Lifetimes

A star's lifetime depends on its mass, which is determined by the ''genetics'' (size and dynamics) of the parent gaseous cloud from which it contracted. In the animal kingdom, generally, the larger the animal, the longer the lifetime. In the stellar kingdom, however, the opposite holds true—giant massive stars burn out quickly and have short lifetimes, while extremely small dwarf stars live the longest.

Stellar lifetimes range from 1 million years for the largest supergiants to a possible 100 billion years for the smallest dwarf stars—a ratio of 100,000 to 1. The animal kingdom offers no comparable ratio. Even the longest-lived tortoises—about 200 years old—live only 2,900 times longer than the common housefly, whose average life span is about 25 days. Comparing stellar lifetimes of 1 million and 100 billion years is like comparing a single afternoon to an entire average life span of 72 years.

New Meaning for the Little Dog

Procyon, the Little Dog Star, lies in the constellation Canis Minor and during the winter months rises in the east from 15 to 20 minutes before the Great Dog Star, brilliant Sirius. Procyon is similar to the Sun in some ways, although it is actually a great deal brighter. Observers can find it by looking for a yellowish star northeast of Sirius slightly more than 20 degrees—about the same apparent distance from Sirius as is famous Betelgeuse in Orion.

This star is impressive not so much for its visual aspect, which is diminished by great Sirius, but by the knowledge that it may outlive our Sun. If you want to see a star that may shine on after the death of our Sun about 5 billion years from now, go look at Procyon on a clear winter night. Somehow the night will seem colder.

STARBIRTH

The Dark Before Light

Interstellar gas clouds, dark or illuminated, are the stellar blood for gestating stars. As these immense, tenuous nebulae condense, they become dark irregular shapes against the back-

An x-ray image of the nearest star system to our Sun. It would take 850 thousand years for an Apollo spacecraft, at its Earth-Moon speed, to reach Alpha Centauri. Courtesy Einstein Observatory/Harvard-Smithsonian Center for Astrophysics.

The famous Pleiades star cluster contains young, hot stars no more than 50 million years old. They lit up about the time the first grasses on Earth began to grow. Courtesy Mount Wilson and Las Campanas Observatories, Carnegie Institution of Washington.

ground starlight and are called *Bok globules,* after the American astronomer Bart J. Bok. These globules represent the first time that condensing protostars become visible as their temperatures and pressures built up over millions of years to the threshold where nuclear fusion begins—and stars are born. The largest Bok globules could contain 312 of our solar systems, with diameters equal to over 12,000 times the Earth's orbit around the Sun.

A Stellar Frisbee

Collapsing clouds of interstellar matter eventually form stars, and the condensing matter may also form planetary systems at the same time. A star going through this evolutionary stage is called a *disk star,* since the planetary matter forms a rotating disk around the central star.

In 1977 scientists discovered what they believed to be the first disk star—MWC 349, about 8,000 light-years away from the Earth in the constellation Cygnus—by making infrared observations that could ''see'' through the thick veil of dust surrounding it. The luminous disk of hot, glowing gas has a diameter 20 times that of the central star, but a nonluminous portion extends much farther and may already contain the newly formed outer planets of the system. The central sun's diameter is calculated at 10 times our Sun's—about 8.6 million miles (13.8 million kilometers). This means that just the visible, glowing portion of the disk is about 172 million miles (277 million kilometers) in diameter—almost twice the distance of the Earth from the Sun. Since the central star has a mass about 30 times our Sun, its stellar evolution will be extremely rapid in comparison, and intelligent life may not have time to evolve. If intelligent planetary life did evolve, it would have to survive its star's demise—which will first explode as a supernova and then collapse as a black hole. And that is only about 60,000 years off, since MWC 349's nuclear fires began burning just 10,000 years ago.

Newborn Stars

Astronomers believe that the class of T Tauri stars, named after the first to be discovered in the constellation Taurus, are just-born stars. These stars are always associated with gigantic gaseous nebulae from which they condensed. Their luminosities vary erratically, probably because they are still growing and accreting material before they reach the stable main-sequence state. The star T Tauri is expected to evolve to the main sequence about 10 million years from now, when a long succession of earthquakes along the San Andreas fault will have moved Los Angeles more or less to the latitude of San Francisco.

Stellar Gestation

The gestation period for a star about the size of our Sun—from its embryonic protostar contraction to its nuclear-fusion birth—is about 50 million years, 67 million times longer than human gestation, the evolution of which took about 3 billion years.

A Creative Cloud

The great Orion Nebula, a cloud from which stars are forming, is at least 30 light-years in diameter—177 trillion miles (285 trillion kilometers). This is enough space to place 20,000 of our solar systems end to end. If an edge of this vast

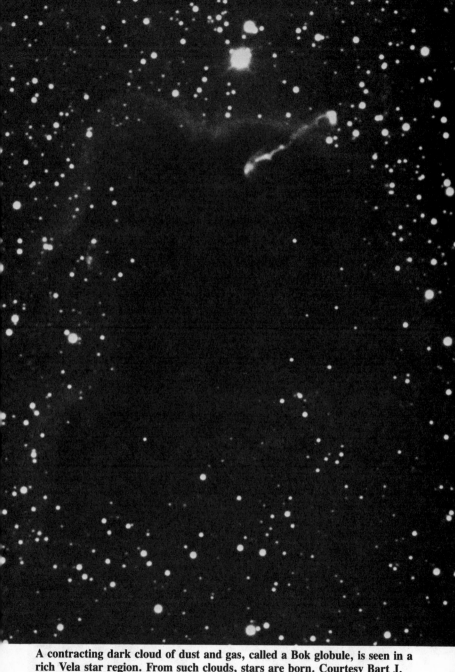

A contracting dark cloud of dust and gas, called a Bok globule, is seen in a rich Vela star region. From such clouds, stars are born. Courtesy Bart J. Bok, Steward Observatory.

(Top) The Great Orion Nebula, one of the most beautiful starforming clouds in the Galaxy. (Bottom) In this stellar womb are four stars in a cluster called Trapezium that are still contracting. Courtesy Mount Wilson and Las Campanas Observatories, Carnegie Institution of Washington; U.S. Naval Observatory.

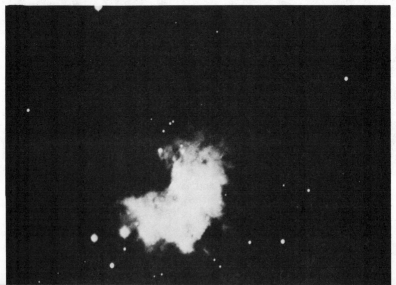

glowing cloud were placed near our Sun, it would stretch 7 times farther than our nearest triple star group, Alpha Centauri. If the distance between the Earth and Sun (an astronomical unit) were represented by 1 inch, the great Orion Nebula would be 12.6 miles (20.3 kilometers) in diameter. This vast, creative cloud has enough material for 10,000 stars like our Sun; in fact, they have already started to condense, and the birthrate continues to be high.

An Embryonic Star

About 1,600 light-years away, in the glitter of Orion's sword, lies the beautiful Orion cloud of glowing green gas and dust—the nearest and brightest nebula to our Sun and planets. This cosmic cloud is illuminated by Theta Orionis, once considered the central star in the sword of Orion but now known to be a cluster of four stars called the Trapezium. These young variable stars are still forming, gravitationally contracting out of the gas and dust cloud. In fact, the whole Orion Nebula is considered a stellar womb, where new stars are constantly being born. Many of these so-called protostars are not visible to optical astronomers because of the thick veil of gas, but infrared telescopes, "seeing" beyond the red end of the visible spectrum, can learn much about them.

Becklin's star (a Becklin-Neugebauer object) is one such intense infrared source—a protostar with a mass about 10 times the Sun and 1,000 times brighter, keeping its cool at 327 degrees Celsius (620 degrees Fahrenheit, not much more than an oven broil). Becklin's star was probably stirred to life about 10,000 years ago—when Neolithic man on Earth began to live in villages and cultivate plants for food. Millions of years from now, perhaps, Becklin's star may feed the greenery of a planet Orionis, and after the green may come other life.

☆   ☆   ☆   ☆   ☆

STELLAR PROFILES

The Brightest Star Ever Recorded

Eta Carinae is an unusual variable star embedded in a cloud of luminous gas (the Keyhole Nebula), and astronomers have had difficulty classifying it. Is it a slow nova or a supernova? Explosive stars usually show a sharp rise in brilliance and then fade slowly over time, but Eta Carinae was bright for over 100 years before it flashed its record brilliance in 1843. During this peak, Eta Carinae was the second brightest star in the sky—Sirius held on to its title even then. But while Sirius was a mere 9 light-years away, Eta Carinae was over 6,000 light-years away from our solar system. The 1843 star flash has been estimated at several million times more luminous than the Sun—perhaps as much as 4 million times—making Eta Carinae the most brilliant and luminous star ever recorded. If it did at any point equal 4 million Suns, it could have been observed through a powerful telescope (such as the 200-inch Mount Palomar reflector) from over 450 million light-years away, 200 times the distance to the Andromeda Galaxy.

What price glory? Just 25 years later, in 1868, this tremendous star was no longer visible to the human eye. Some astronomers believe that Eta Carinae exploded, leaving only its remnants expanding throughout the thick gas and dust of the Keyhole Nebula. It would seem that the brighter they are, the sooner they fizzle.

Barely Lit

One of the faintest stars known, van Biesbroeck's star, is about 20 light-years away and is in the cool M-class group of stars—a red dwarf. It is 570,000 times dimmer than our Sun, 700 times brighter than the planet Jupiter, and if it were in our Sun's position, it would shine on Earth only a bit brighter than a full Moon. At its actual distance, van Biesbroeck's star is, of course, invisible to the naked eye, because to see it would be like seeing a 60-watt lightbulb at a distance of 34,000 miles (55,000 kilometers)—a distance equal to over 4 times the Earth's diameter.

A Brillant Star, Far, Far Away

The star S Doradus is a bright, variable supergiant located in the Large Magellanic Cloud, a satellite galaxy to our own Milky Way almost 200,000 light-years away. It is one of the brightest stars known in the Universe—its average luminosity is 500,000 times brighter than our Sun, and on occasion, because of its variable cycle, it has exceeded the Sun's brightness by over 1 million times. If S Doradus were located over 700 times farther away from Earth than our Sun—about 65 billion miles (104 billion kilometers)—it would still provide our planet with as much energy as we get from the Sun.

A Sun-Dwarfing Star

One of the largest stars known in the Universe is the red giant VV Cephei, located about 3,900 light-years from Earth. Its diameter is 1,900 times greater than the Sun. If the Sun were represented by a chick-pea, VV Cephei would be a hot-air balloon 39 feet (12 meters) in diameter. If this star were centered on our solar system, it would fill space as far out as Saturn, which is over 900 million miles (1,450 million kilometers) from the Sun.

The Largest Star?

More than 50 percent of all stars are in binary systems, where two stars move through space together and influence one another gravitationally. Such star pairs represent the most common stellar nuclear family, and, depending on the Earth's perspective relative to the twosome, one star often eclipses the other, cutting off all or a proportion of its light. These stars are known as *eclipsing variables,* and one of the most mysterious of this type is Epsilon Aurigae and its invisible companion, which revolve around a common center of gravity once every 27 years. The invisible companion substantially eclipses the bright supergiant optical star for about one year once every revolution.

The mysterious companion star of Epsilon Aurigae has never been observed optically or spectroscopically. While some astronomers believe that the unseen companion may be a smaller star surrounded by a great cloud of gas or encircled by a flat disk of gas that is viewed edge-on, the traditional interpretation says it is an extremely low-density supergiant whose temperature is so low, about 1200 degrees Celsius (2200 degrees Fahrenheit), that no visible light can be seen from Earth. If this view is correct, the invisible companion of Epsilon Aurigae may be the largest star known—2,800 times the diameter of the Sun (and only one billionth of the density), or 3.5 billion miles (5.7 billion kilometers). If centered on our solar system, the edge of this stellar colossus would fill the orbit of faraway Uranus, and its volume could comfortably accommodate 22 billion Suns.

The Heaviest Giants

Plaskett's star, about 2,700 light-years away in the constellation Monoceros, is really two giant 0-type stars, separated by a mere 50 million miles (80 million kilometers), that orbit a common center of gravity every 14 days. These two stars are considered by most astronomers (including their discoverer, Canadian astronomer John S. Plaskett) to be the most massive stellar system yet discovered in the Galaxy, but exact values have proved difficult to obtain because of the turbulent masses of gas swirling around and between them at such a close distance. Recent estimates (less than earlier ones) give the larger star a mass of 60 times the Sun, with a mass for its companion of 40 times the Sun. A star weighing 60 times the Sun would equal more than 20 million Earths, 163 million Moons, or 160 billion asteroids the size of our largest, Ceres.

The Fastest Star in the South . . . and North and East and West

A dim red-dwarf star, Barnard's star, with a diameter of about one sixth the Sun's (140,000 miles; 225,000 kilometers), is traveling faster than any other star as viewed from Earth. Barnard's star, named after astronomer Edward Emerson Barnard of Yerkes Observatory, who discovered it in 1916, is the second closest star to the Earth (about 6 light-years), if the triple Alpha Centauri system is counted as one. Its mass is only 16 percent of the Sun, and its temperature is a cool 2900 degrees Celsius (5250 degrees Fahrenheit).

This speedy dwarf is located in the constellation Ophiuchus, and a telescope must be used to see it in the southern skies in August and September. How fast is a fast star? No doubt there are plenty of faster moving objects in the Universe, but because they are so distant, they do not move across the sky like Barnard's star—which can travel two full-Moon diameters about every 350 years. This means that this red dwarf is moving toward the Earth at just over 24,000 miles (39,000 kilometers) per hour. In another 10,000 years, Barnard's star will have traveled more than 2 light-years toward the Earth, and it will then hold the title as the closest star to our solar system.

A Steaming Star?

The so-called Garnet star in the constellation Cepheus, named by Sir William Herschel because of its unique red color, is a red giant similar to Betelgeuse in Orion and is considered by most observers to be the reddest star visible in the northern hemisphere. But its spectacular color, which varies dramatically depending on the telescope and optical power used, is not its most intriguing feature. The Garnet star (Mu Cephei) is one of the few stars known whose spectrum shows hot water vapor—in other words, steam!

No fully satisfactory explanation has been found, but some astronomers believe it may be due to the star's turbulent atmosphere and swirling high velocities. If a red giant's outer atmosphere is relatively cool, a steaming star is certainly possible.

The Old Ones

The oldest known star cluster in our Galaxy is located in the constellation Cepheus, fairly close to Polaris, the North Star, and is therefore observable throughout the year. Designated in the *New General Catalogue* as No. 188, this ancient cluster is about 5,000 light-years away and 1,800 light-years above the galactic plane.

No young, white-blue, brilliant stars exist any longer in this star group; they died billions of years ago. The brightest and youngest stars remaining are yellow giant stars, which are

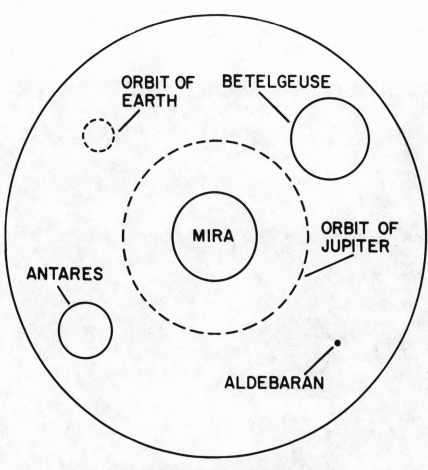

ORBIT OF EARTH

BETELGEUSE

MIRA

ORBIT OF JUPITER

ANTARES

ALDEBARAN

EPSILON AURIGAE
2,300,000,000 MILES

Comparative sizes of giant and supergiant stars. Epsilon Aurigae may be the largest known star in the Universe. If its estimated size is correct, it could accommodate 22 billion suns. Courtesy Science Graphics, copyright © 1982.

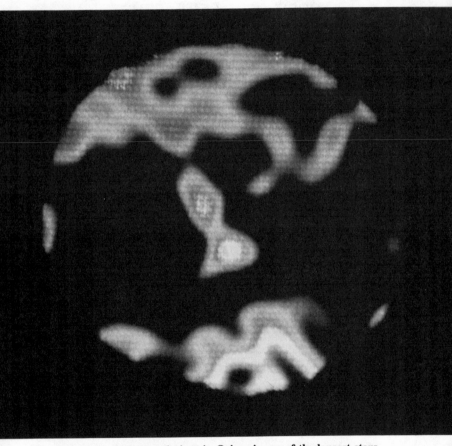

Betelgeuse, the famous red giant in Orion, is one of the largest stars known, with a maxium diameter of about 900 times that of our Sun. Courtesy Kitt Peak National Observatory.

in the 1- to 1.5-billion-year age range. Some of the oldest surviving stars in our Galaxy are likely to be found in cluster NGC 188, which has an estimated age of 11 or 12 billion years. This would be the age of the surviving oldest stars, most likely G-6 main-sequence stars with 5 percent less mass than our Sun, that are close to evolving off the main sequence.

A 12-billion-year-old star would have blazed its youthful glory about 7 billion years before our Sun and planets formed, and about 6 billion years before the supernova exploded that probably triggered the condensation of the metal-rich interstellar cloud that eventually resulted in the Sun, the planets, and life on Earth, which in fact provided the very atoms of our bodies.

Armpit of the Giant

Betelgeuse (Alpha Orionis), located 520 light-years away in the constellation Orion, is a red pulsating supergiant, and one of the most famous great stars in the sky. This red giant has been well known for thousands of years by all cultures—its name in Arabic means ''Armpit of the Giant.''

The eleventh brightest star in the sky, Betelgeuse is one of the largest stars known and probably the largest star that can be seen without a telescope or binoculars. While its size varies because of irregular pulsations, its maximum diameter is about 800 million miles (1,300 million kilometers), which would fill our solar system from the Sun to almost the orbit of Saturn. This size is over 900 times the diameter of the Sun and over 160 million times its volume, but during its pulsation cycle, it contracts all the way down to 550 times the diameter of the Sun—480 million miles; 772 million kilometers—which amounts to a difference larger than the Earth's orbit around the Sun.

Betelgeuse's Halo

In the late 1970s, for the first time ever, matter was photographed leaving the red-giant star Betelgeuse—an immense stellar wind ejected from its surface. It is this constant gaseous flow from most stars that seeds the interstellar medium and makes possible the birth of new second-generation stars that contain the heavier elements essential to the evolution of life. In just 1 million years at its current rate, Betelgeuse would expel an amount equal to the Sun's mass; in just 3 years, this star stuff would equal the mass of the Earth. Betelgeuse's potassium gas cloud is 400 times larger than our solar system, 188 million times larger than the Earth's equatorial diameter— about 1.5 trillion miles (2.4 trillion kilometers). Proportionately, Betelgeuse's gaseous halo is like a candle flame having a 100-foot (30.5-meter) halo.

The Rival of Mars

The ancients called Antares (Alpha Scorpii) "the rival of Mars" because they both were uncommonly red in color and they sometimes shared the same portion of the sky. This tremendous star is another red supergiant like Betelgeuse, with many similar characteristics. Its diameter is about 700 times that of the Sun (600 million miles; 965 million kilometers), as compared to the 900 times for Betelgeuse. Both red supergiants have the same temperatures—a stellar cool of about 2800 degrees Celsius (5100 degrees Fahrenheit). All red giants, in fact, share these relatively cool temperatures as they expand and evolve to this advanced state of their life spans after the hydrogen in their cores has been used up.

The starshine of Antares is 9,000 times brighter than our Sun (Betelgeuse is 14,000 times brighter). While its mass is only about 15 times that of the Sun (20 times for Betelgeuse), like all supergiants, its volume is gargantuan and its density is tenuous—less than one millionth of our Sun's density, which is only 1.4 times the density of water. This makes Antares and similar stars "red-hot vacuums." The Earth could not exist if Antares were in the Sun's position; the Earth would have to be far beyond Pluto for it to remain a planet.

☆   ☆   ☆   ☆   ☆

THE VIOLENT STARS

The Flashes Are Coming

About the time our apelike ancestors split off from the line leading to the true apes, 15 or 16 million years ago, a giant star exploded near the distant galaxy NGC 5253. The light from this extragalactic supernova event finally reached Earth in May 1972 and was first discovered by astronomer Charles T. Kowal at Palomar, who was involved in a systematic supernovae search. This supernova outshone the light of its entire galaxy by 10 times—an estimated luminosity almost equaling 13 billion suns. Galaxy NGC 5253 has probably had about 50,000 more recent supernovae in the last 15 million years, and their explosive flashes are now racing toward Earth for some future astronomer-astronaut to detect.

Saved by Space

The average space between the stars is a blessing for life. If stars were closer together, nova and supernova explosions would endanger or kill most life forms living within 2 or 3 light-years. Invisible high-energy radiation would be the killer.

The Brightest Debris

Chinese, Japanese, and Arab astronomers witnessed in A.D. 1054 the most famous supernova of the seven recorded in our Galaxy. The Chinese called it a "guest star," and, at a calculated distance of about 6,300 light-years, its novel brilliance no doubt seemed pleasant enough. The fact is that this supernova in the constellation Taurus was so violent that it blazed with a peak intensity equal to about 500 million suns. From Earth, it was brighter than Venus and was visible during daylight for over three weeks. The ancient astronomers were unknowingly looking back through time and space to an event which actually occurred about 6,300 years before—about 5200 B.C., when the Sumerians had established themselves in lower Mesopotamia, corn was cultivated in Mexico, and the boomerang was used by native Australians.

The 1054 supernova is more famous for what it left behind than for its original appearance in the eleventh century. The debris from this great stellar explosion forms the Crab nebula, named by the Irish astronomer Lord Rosse, who saw a crab in his sketch of it; it is a vast expanding cloud of hot, energized gas, and one of the most studied objects in all astronomy.

The Crab nebula (*New General Catalogue* No. 1952) that astronomers observe today has a total luminosity of about 30,000 suns and is the brightest supernova remnant in the sky. It emits powerful radiation at all wavelengths and is one of the four strongest radio sources (named Taurus A) known to astronomers. That the Crab nebula was still so bright and powerful 900 years after its light image reached Earth was puzzling to astronomers—until the remnant core of the exploded star, a pulsar-neutron star, was discovered at its center. This star's tremendous ultraviolet energy heats the expanding gas and gives us the wonderful and spectacular gas cloud we observe,

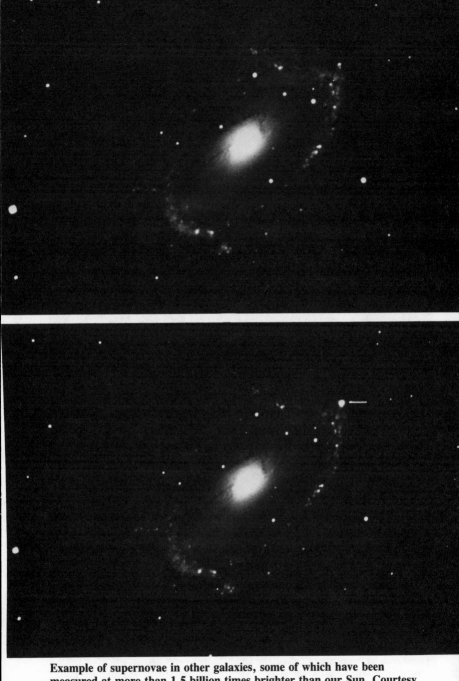

Example of supernovae in other galaxies, some of which have been measured at more than 1.5 billion times brighter than our Sun. Courtesy Palomar Observatory, California Institute of Technology.

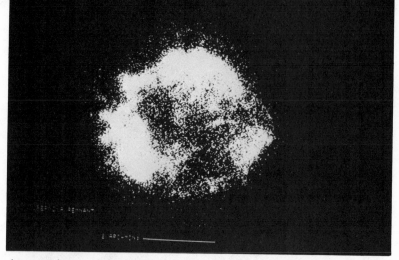

An x-ray image of the remnants of Tycho's star (Cassiopeia A)—the lost and found supernova. Courtesy Einstein Observatory/Harvard-Smithsonian Center for Astrophysics.

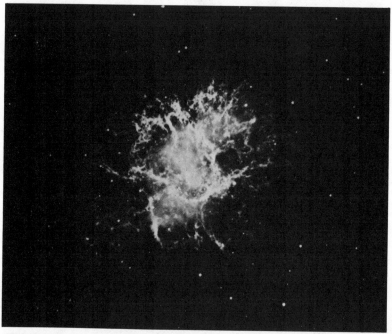

The most famous supernova of our Galaxy took place in 1054 A.D. Its expanding remnants are called the Crab Nebula—an astronomical spectacular with a surprise at its center. Courtesy Mount Wilson and Las Campanas Observatories, Carnegie Institution of Washington.

an expanding cloud that travels 50 million miles (80 million kilometers) every day and has a diameter of 6 light-years or 35 trillion miles (56 trillion kilometers), which is over 10,000 times the distance from the Sun to the farthest planet, Pluto. If the Crab nebula continues to expand at its present rate, without any interstellar obstacles, its vague remains, already more tenuous than one trillionth the density of ordinary air, will reach the Earth in 2.04 million years.

Lost and Found Supernova

Over 400 years after Tycho Brahe observed the famous supernova in the constellation Cassiopeia that would be named after him (Tycho's star), astronomy knows: How far away it was (about 10,000 light-years); how much brighter than the Sun it was (300 million times brighter); how large the remnant supernova cloud is (about 20 light-years or 35,000 times the size of our solar system); how fast the expanding remnant cloud is traveling (over 20 million miles or 32 million kilometers per hour); and how hot the remnant cloud is (about 1 million degrees Celsius). Tycho Brahe observed the event before the first telescope was invented, and in 1574 his "miracle" star became invisible. Thanks to Tycho's notes, however, two radio astronomers found an intense radio source in 1952 that proved to be its remnants. The lost supernova was found again 378 years later.

Two Tugging Giants

Two tear-shaped stars tugging at one another—that is a simplified description of the Beta Lyrae binary system. Beta

Lyrae has been observed, investigated, and analyzed more than any other star, with the exception of our Sun, and a doctoral candidate will no doubt someday write a dissertation just on the evolution of computer models that have tried to capture the reality of this mysterious duo. These tugging giants head a class of Beta Lyrae–type stars of which a few hundred are known.

Both stars are larger than the Sun—the primary about 19 Sun diameters and the smaller about 15. They revolve around a common center of gravity and eclipse every 12.9 days; thus they are called eclipsing variables. But the amazing fact is not that Beta Lyrae is an eclipsing two-star system, which is quite common in the Universe, but that the stars are actually touching and influencing one another. Even though they are separated by about 22 million miles (35 million kilometers), this distance shrinks to insignificance when star diameters of a few million miles less are brought close together. They are so close, in fact, that their atmospheres mix together and an actual transfer of mass takes place, a vast stream of gas that flows from the large star to the smaller companion at a velocity of 648,000 miles (1,042,000 kilometers) per hour, traveling the great distance in about 31 hours. The smaller star receiving the gigantic gas stream is faint and mysterious, and some astronomers suggest that much of its mass (which is twice as much as the larger star) is contained in a vast cloud or disk of gas. As the gas stream continues to flow, the stars move closer together, which in turn creates a greater flow of gas. Eventually the masses will become equal, and the smaller star will probably become the larger, dominant one. The mass flow will reverse itself—an exchange of stellar roles!

The amount of mass flowing between the Beta Lyrae stars in the estimated 63,500-year flow period amounts to about 200,000 Earths.

A Christmas Nova

Nova Herculis, a normal nova, brightened to its maximum just in time for Christmas 1934—a maximum brightness measured to be 400,000 times its pre-nova light. At an estimated distance of 1,200 light-years, Nova Herculis outshone our Sun by about 65,000 times.

Mounting evidence indicates that most novae stars are hot blue or white dwarfs in a two-star (binary) system and that the tremendous gravitational forces and mass transfer between the two stars play a role in novae explosions. Nova Herculis has a companion star with an extremely short orbital period—about 4.5 hours—which causes the pre-nova dwarf to pulsate and which may trigger it to go nova.

Cosmic Gas Rings

Some stars blow rings—not of smoke but of gas—and the great astronomer William Herschel called them "planetary nebulae" in 1785 because they appeared as disks, similar to planets, through his early, not-too-powerful telescope. They have nothing to do with planets, however, only with stars, and it is believed these vast rings of luminous gas are thrown off by stars over a certain size (perhaps 1.5 solar masses) when they evolve to the red-giant stage near the end of their lifetimes. This way the star rids itself of perhaps 10 percent of its mass before it enters the white-dwarf state to eventually fade away.

The stars in the center of the great gas rings are the hottest known. Some have been measured at over 150,000 degrees Celsius—hotter even than the Wolf-Rayet stars (see next entry) from which they may have evolved.

The brightest and nearest planetary nebula, the Helical nebula (NGC 7293) in the constellation Aquarius, is also one of the largest, with about half the apparent width of the full Moon. The true diameter of this nebula, at an estimated distance of 450 light-years, is about 1.75 light-years, over 110,000 times the distance from the Earth to the Sun—enough space to line up over 2,700 solar systems along it. But this vast luminous gas ring probably has no more mass than one tenth of the Sun.

A Rare Stellar Breed

Two Paris Observatory astronomers discovered a spectacular class of stars in 1867 that are among the hottest and most violent stars known. Wolf-Rayet stars (Type W)—named after Charles J. E. Wolf and Georges A. P. Rayet, the two discoverers—are extremely luminous and have surface temperatures reaching 100,000 degrees Celsius in some cases—over 17 times our Sun's surface temperature (about 5700 degrees Celsius).

Wolf-Rayet stars are blasting away tremendous amounts of their outer layers into surrounding space at speeds of up to 6.7 million miles (10.8 million kilometers) per hour or 122,000 times faster than the U.S. 55-miles-per-hour highway speed limit. These velocities of expanding gas equal those of stellar novae explosions, but while novae explode only once or twice in their lifetimes, the hot blue Wolf-Rayet stars continually eject such huge amounts of gas. Since they average twice the mass of the Sun, even though their turbulent atmospheres are much larger, they can lose their complete mass in just a few million years; thus they are short-lived and rare (only 200 known, 150 of which are in our Galaxy).

Where do Wolf-Rayet stars fit into stellar evolution? Astronomers are not sure. Some experts believe they represent the cores of red supergiants after the outer layers have blown

Nova Herculis, one of the brightest novae of this century, remained brilliant for about 100 days. Courtesy Yerkes Observatory.

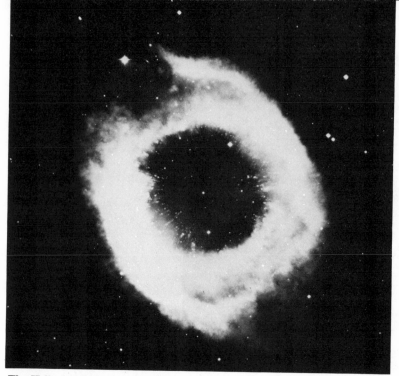

The Helical Nebula (NGC 7293), a vast luminous gas ring, marks a star's change of life—from red giant to white dwarf. Courtesy Palomar Observatory, California Institute of Technology.

Violent Wolf-Rayet stars blast off their outer layers at tremendous velocities. Some astronomers believe they represent the transition from red supergiant to white dwarf. Courtesy D.F. Malin. Anglo-Australian Telescope Board, © 1979.

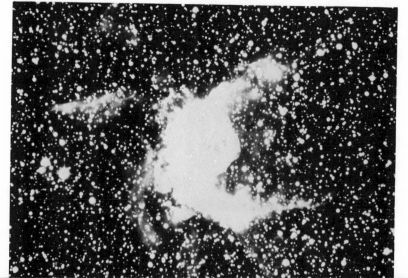

off. This view is supported by the fact that some Wolf-Rayet stars have been found in the centers of great luminous gas rings called planetary nebulae, which may be the blown-away outer layers of the dying red giants. The remaining core would eventually become a white dwarf and slowly die away. Other experts believe that these oddly behaving stars all have close massive companion stars that gravitationally influence their behavior. The star Gamma Velorum, about 500 light-years from our solar system, is a classic example of the hot, violent, and still somewhat mysterious Wolf-Rayet star. If our Sun lost as much mass as these ultra-hot stars, it would only have had a 1-million-year life span rather than its expected 10 billion years.

☆   ☆   ☆   ☆   ☆
WHITE DWARFS

Heavy-handed

Sirius B, almost 9 light-years away from us, is an exceptional star—a white dwarf, the first of its class to be discovered. While it is just slightly smaller than the Earth, it contains so much matter that it weighs nearly as much as the Sun. A handful of its matter would weigh about 500 tons.

Diamonds in the Sky

Some white-dwarf stars—at certain ages, temperatures, and densities—are probably composed of a crystalline form of carbon known as one of the hardest substances on Earth—diamonds. No doubt, they would be the largest diamonds in the Universe, about the size of the Earth.

☆   ☆   ☆   ☆   ☆
NEUTRON STARS

Firmly Packed

After a supernova explosion, the star's compressed core remains—a neutron star (pulsar). A pinhead's worth of neutron star stuff would weigh about 1 million tons—over 10 times the weight of the aircraft carrier U.S.S. *Nimitz*.

Popcorn Energy

If a piece of popcorn were dropped on a neutron star, it would produce as much energy as a World War II atomic bomb.

The Middle-Class Degenerates

If a dying star is more than 1.44 times the mass of the Sun (the limit for white-dwarf formation) but less than 3.2 times the Sun's mass, it will blast off its outer layers in a supernova explosion and become a neutron star (pulsar). This dying star core collapses to such an extreme state that it can be compared to a gigantic atomic nucleus, where the negatively charged electrons are fused onto the positively charged protons and become no-charge neutrons, squeezed so tightly together that they almost touch one another, which, in turn, halts the collapse. A stable neutron star is thus formed, which holds a stellar territory between the white dwarfs and the black holes as dying stars. Most of the 300-plus neutron stars discovered in our Galaxy have been found by radio astronomers as pulsars—rapidly spinning neutron stars that throb out x-ray and other radiations as they rotate.

The small white dwarf, Sirius B, visible to the left of its companion, and which appears to be a droplet attached to it, is smaller than the Earth but weighs as much as the Sun. Lick Observatory photograph.

SUN
(864,000 MILES)

WHITE DWARF
(~10,000 MILES)

RED GIANT
(200,000,000 MILES)

**A white dwarf compared to the relative sizes of a red giant and the Sun.
Courtesy Science Graphics, copyright © 1982.**

NEUTRON STAR
(~10 MILES)

BLACK HOLE
(~3 MILES)

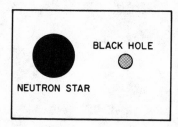

BLACK HOLE

NEUTRON STAR

WHITE DWARF
(~10,000 MILES)

**A white dwarf compared to the relative sizes of a neutron star and a black
hole.**

If a large neutron star, about 12 miles (20 kilometers) in diameter, were in the Sun's core, it would not cause a solar catastrophe. It rather could serve as a source for gravitational energy generation and keep the Sun shining stably at its present rate for much longer than the projected future of 5 billion years—for another 3.2 trillion years, in fact.

Too Heavy to Hold

A 9-pound baby would weigh 90 billion pounds on the surface of a neutron star.

A Colossal Midget

Even though a small neutron star weighs over 400,000 times the Earth, its diameter is about 1,300 times smaller than the Earth's—or 6 miles (10 kilometers), an on-Earth distance a person can comfortably walk in two and a half hours. If the Earth were compressed to the density of a neutron star, its diameter would be about 440 feet (134 meters), and it could easily fit into the Astrodome.

☆　☆　☆　☆　☆
BLACK HOLES

The Best-Bet Black Hole

Cygnus X-1 was one of the first cosmic x-ray sources to be discovered, back in 1962, seven years before the United States landed men on the Moon. Instruments were aboard a short suborbital Aerobee rocket flight, since this was before the time of orbiting x-ray satellites and observatories, which would begin in the early 1970s. More than 5 years passed before astronomers could finely tune the location of the x-ray source (with the help of a new, unexpected radio source) and find a visible star that was located near it. Now astronomers had what they wanted: an optical star whose light (spectrum) they could measure and analyze. Research became more intense, and the star's light revealed its secrets, one of which was that its motions were being influenced by an invisible nearby companion. The first possible black hole was discovered—Cygnus X-1.

The visible star turned out to be in the extremely hot (close to 30,000 degrees Celsius) B class—a giant blue hot star designated HDE 226868* that was about 25 times the mass of the Sun. The x-ray intensity of Cygnus X-1 was found to fluctuate rapidly on a time scale of one twentieth of a second, and this fact told astronomers that the invisible x-ray object is extremely compact (less than 100 miles or 161 kilometers in diameter), even though its mass is at least 10 times the Sun, perhaps as great as 20 times. Since this mass is well above the limit for a stable neutron star (about 3.2 solar masses), Cygnus X-1, thanks to its visible companion star, is telling us it is a black hole.

Twenty years after its first discovery, astronomers have painstakingly pieced together the following model of the mysterious stellar system that is about 8,000 light-years away (our Sun is just over 8 light-minutes away). The visible blue supergiant star is losing a portion of its mass to the black hole as a great hot stream of gas flows between the two bodies. This gas stream does not flow directly into the black hole but builds up and spirals around it before working its way inward toward the boundary of no return—the event horizon. This accumulation

---

*Star number 226,868 in one of the largest star catalogues ever made, compiled at Harvard and named after astronomer Henry Draper, the first American to study stellar spectra.

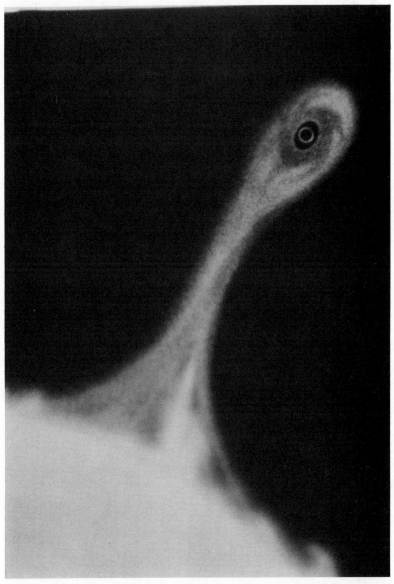

Cygnus X-1, the first black hole, pulls hot gas from its companion star into its gravitational well in this artist's rendering. From the film, "The Universe," courtesy NASA.

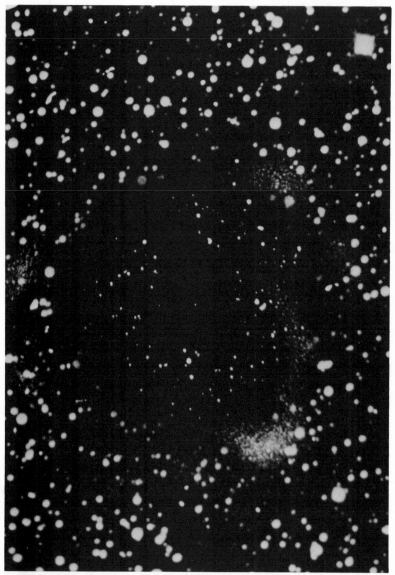

A large black hole can gobble up enough matter each second to equal over three Earths. From the film "Birth and Death of a Star," copyright © 1974 American Institute of Physics.

of spiraling gas is called the *accretion disk,* about 124 miles (200 kilometers) in diameter for Cygnus X-1. It is this gas which emits the flickering x-rays picked up and analyzed by Earth-based cosmic detectives. This means that an invisible black hole "dwarf" (in other words, a collapsed star), whose girth (event-horizon diameter) is less than 100 miles (161 kilometers), is tearing off and gobbling up the stellar flesh of a blue supergiant about 6,000 times brighter and 25 times more massive than our Sun!

A Well-Fed Black Hole

If a black hole formed from a collapsed star is well fed with a gas stream from a nearby main-sequence star or from a rich interstellar medium, perhaps left behind from an earlier supernova explosion, it will grow larger. As the black hole grows, so does its gravitational influence. Star-collapsed black holes range from about 3 to 30 solar masses, but with continuous feeding of matter, there is no theoretical limit to their size. In fact, some astronomers believe that many elliptical galaxies have supermassive black holes at their centers that could be 10 billion times larger than the mass of our Sun—equal in mass to an entire small galaxy. If the largest possible star-collapsed black hole of about 30 solar masses could continuously feed off accreting matter, it could gobble up the equivalent matter of 3.52 Earths each second, or 111 million Earths every year, if each one stood in line and waited its turn.

The Last Wink

Black holes are the gravitational ghosts of once-large stars (over 3.2 times the Sun's mass) that have collapsed to infinity and have left the space-time of our Universe. White dwarfs and neutron stars, smaller than 3.2 solar masses, halt their collapse because the subatomic electrons, protons, and neutrons create a holding pressure, at certain subatomic densities, determined by the star's original mass. Not so with larger stars: They collapse to an incomprehensible black-hole state of infinite density and zero volume. This state is called *singularity,* a never-never reality where even relativity theory breaks down and tells theorists nothing. When one of these stars collapses past a certain diameter (about 5.6 miles or 9 kilometers for one with 3 solar masses), it winks out, allowing no light, no radiation whatsoever, to escape. Goodbye, star, forever. The single thing scientists can know about the object that formed the black hole is how massive it was; that is all. Only the gravity remains. The black hole may be bottomless, or it may be a tunnel leading to another Universe. There is no way for science to know, and all speculation about wormholes and white holes is exactly that—speculation.

How long does a black hole take to collapse? Less than a second—about as much time as it takes a balloon to pop or for 8 million blood cells to die within one human adult.

The Last Collapse

When a star at least 3 times as massive as our Sun collapses, it forms a black hole, where only the star's gravitational force is left behind. The largest single-star-formed black holes are believed to have gravitational boundaries (event horizons) of 93 miles (150 kilometers) in diameter, formed when giant stars, with about 50 times the mass of our Sun, experience their final death throes and go through gravitational collapse.

The largest stellar-formed black holes may be at the centers of active galaxies. Their masses could be equal to 10 billion Sun masses, and their event horizons could be over 18 billion miles (29 billion kilometers) in diameter—almost 2.5 times the diameter of our solar system.

Finally, there is the Universe. If its mass exceeds a critical value, then it can be considered a black hole which will someday collapse.

Black-Hole Energy

There have been some way-out speculations on far-future ways that humankind can exploit the immense energies surrounding near black holes—assuming, of course, that we get our starship fleets launched in the next few thousand years and that technology in the next 100,000 years continues to advance at the current explosive rates.

If future humankind could erect a huge steel flywheel at just the right distance away from a black hole, the rotating space would also cause the giant flywheel to rotate, and colossal amounts of energy would be generated for the nearby civilization. Such black-hole-produced energy could supply the Earth's current level of consumption for billions and billions of years—but by that time the Universe, if it eventually contracts, will no longer exist.

THE GALAXY: OUR COSMIC CAROUSEL

☆ ☆ ☆ ☆ ☆
OUR SPIRAL FROM AFAR

Finding Our Place

We live in a giant spiral star system, the Galaxy (derived from the Greek word *gala,* meaning milk, hence the name Milky Way), which has been one of the splendors of the Earth's night skies since there were eyes to see it. Only in the last half century has its true nature and size been determined by the patient and persistent work of hundreds of astronomers.

Some of the early Greek philosophers, notably Parmenides of Elea, Anaxagoras of Clazomenae, and Democritus of Abdera, had at least reasonable ideas about the nature of the Milky Way: Parmenides, that it was a mixture of dense and rarefied substances; Anaxagoras and Democritus, that it was composed of light from an immense number of stars. No early thinker, however, approached the reality that the Galaxy was a vast rotating wheel composed of billions of stars, one of which was the Sun, situated trillions upon trillions of miles out from the cosmic wheel's hub. That humankind has mapped its position in the Galaxy is an accomplishment greater in scale than if a virus had mapped the entire solar system.

Lost in the Galactic Crowd

The Milky Way Galaxy is one of an estimated 100 billion galaxies in the Universe, and one of 30 billion spiral galaxies. It represents about one trillionth of the Universe, which is like comparing a very small metal screw with the mass of the 100,000-ton ship of which it is a part.

Birth of the Milky Way Galaxy

From the Big Bang came the pristine elements, hydrogen and helium, from which our Galaxy was created. Probably some kind of density variations during the early expanding Universe were amplified in time, once matter became dominant, and caused the condensation of our protogalactic cloud about 12 billion years ago. This great primordial cloud was composed of 75 percent hydrogen and 25 percent helium. No heavier elements existed at this time, nor any stars. Life was impossible.

While much is still unknown about galactic evolution, including the answer to why galaxies are not much smaller or larger than they are, scientists believe our massive protogalaxy was initially in the shape of a great sphere that was rotating. The protocloud thus began gravitational collapse, and when it collapsed to a diameter of about 100,000 light-years, its increasing density caused internal condensations to appear. The great globular star clusters, still existing today, were born from these smaller condensations. They are the oldest objects in our Galaxy, still defining the primordial spherical shape of the protogalactic cloud.

Gravitational collapse continued, leaving the globular clusters behind, and the denser, rotating inner cloud formed into a disk. About 1 billion years later, after several revolutions, spiral arms emerged in the disk where dense gas clouds contracted even further, and clusters of stars were spawned. Many of these original stars burned rapidly and died in convulsive explosions, supernovae, after just a few million years, spewing forth their star stuff. Their demise created new enriched interstellar clouds that contained the first heavy elements, forged in their stellar interiors. From this debris came a second generation of stars. Our Sun was among them, and its birth, along with the planets, had much in common with the Galaxy's but on a vastly smaller scale. A protosolar cloud condensed and gravitationally collapsed, forming a rotating disk of gaseous material that would become the Earth and the other planets.

The elapsed time, from protogalaxy to solar system, was about 7.5 billion years; from solar system to the first microscopic life on Earth, about 1 billion years; from microscopic life to us, about 3.5 billion years. Add it all up and what have you got? About 12 billion years. If these 12 billion years were compressed into 1 year, then humankind appeared on planet Earth about 105 minutes ago, and Galileo made his telescope discoveries about 1 second ago.

The Galaxy's Population

Our Milky Way Galaxy contains at least 200 billion stars, and some estimates go as high as 300 and 400 billion. Imagine a square inch (6.45 square centimeters) with 200 dots inside that represent stars. At the same density, it would take 1 billion square inches (6.4 billion square centimeters), or almost 160 acres, to accommodate the Galaxy's 200 billion stars. This is equal to the area of a standard family farm or a

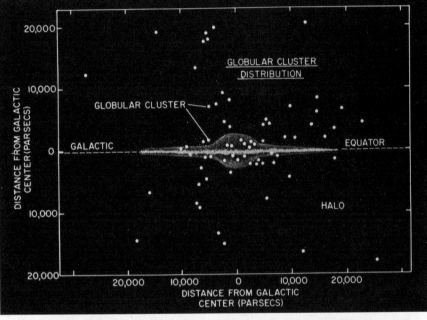

A cross-section view of the Galaxy, showing the great globular star clusters orbiting around it. Courtesy Science Graphics, copyright © 1981.

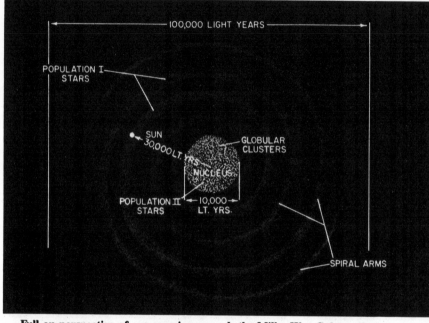

Full-on perspective of our cosmic carousel, the Milky Way Galaxy, that rotates about once every 230 million years. Courtesy Science Graphics, copyright © 1981.

very large suburban shopping center. It is also equal to 23.8 million pages of a book the same size as this one.

If their average spatial relationship in the Galaxy were added to this one-dimensional scale, with 1 star in every 3.26 square inches (21 square centimeters), then the stars would be spread out for 15.3 square miles (1,792 acres), about the area of a small town.

Driving to the Galactic Center

Our Sun and solar system lie about 30,000 light-years from the center of the Galaxy. Even if you were driving a Porsche at 100 miles (161 kilometers) per hour, it would take more time than the entire age of the Galaxy, 12 billion years, to reach the galactic center. Considerably more time, in fact: It would take the Porsche 6.71 million years to travel just 1 of those 30,000 light-years, and a total of 201 billion years to drive to our Galaxy's center.

Scaling the Galaxy

If our solar system (the Sun and nine planets) could fit into a coffee cup, our Galaxy would be the size of North America. Our Galaxy has a diameter of at least 80,000 light-years—that is, 480 million billion miles.

The Milky Way's Temperatures

Since most of the galactic matter is in stars, and since star temperatures and densities rise sharply in their cores, the mean temperature would be some millions of degrees Celsius. Interstellar dust grains, however, far away from the stars, would be

−261 degrees Celsius (−438 degrees Fahrenheit), or one twenty-fifth that of room temperature.

The Finite Galaxy

The estimated mass of our Milky Way Galaxy is equal to about 180 billion suns. Even if this entire mass were converted to energy, it would still not be enough to thrust the dot (period) at the end of this sentence to the speed of light. Theory states that an infinite amount of energy would be required to push the period to the speed of light, and the Galaxy, while immense by human standards, is hardly infinite.

Cosmic Carousel

The Galaxy is a cosmic carousel, and the Earth revolves on it. A galactic year is the time it takes the Galaxy to make one complete revolution—230 million years. One revolution ago, the Dimetrodon (a mammal reptile with a large sail on its back) and its descendants dominated the Earth. The Galaxy is about 52 galactic years old (12 billion years), so it will be another 3 billion years until it reaches the retirement age of 65.

Starship Shortcut

If a starship were launched to travel across the Galaxy, and it averaged a modest cosmic speed of 190,000 miles (305,000 kilometers) per hour, where would it be in 115 million years? It would rendezvous with the Sun and planets on the other side of our Galaxy. Had the starship never left Earth, it would have gotten to its destination at about the same time, but without the adventures along the way—all because of galactic rotation.

Spiraling Within the Spiral

Our entire Galaxy rotates, but the stars, star clusters, gaseous nebulae, and interstellar dust rotate at different speeds (*differential rotation*), depending on their distance from the Galaxy's center. The Sun, planets, and other stars in our galactic neighborhood sail along with their local galactic motion and cover 1.1 million trillion miles (1.8 million trillion kilometers) every 230 million years—a circular path representing 188,000 light-years. Distant Pluto, with an average orbital period around the Sun of about 248 years, completes over 928,000 orbits around the Sun every time the Sun rotates once around the Galaxy—covering a spiraling path within the spiraling Galaxy.

The Empire Flies Off

At the end of the motion picture *The Empire Strikes Back,* as the protagonists tie up the loose ends of their adventures and set the cosmic stage for a sequel, a technicolor spiral galaxy is seen slowly rotating in the background through the spaceship's large portal. The producers probably gave the galaxy visual motion so the audience did not think it was just a still shot; they opted, in other words, for visual effects over astronomical reality. After all, that is what film magic is all about.

In reality, if this spiral were the size of our Galaxy, rotating fast enough for the human eye to detect, its outer parts would be flying around at the impossible velocity of 33 billion times the speed of light!

Revolving Through Time

The Galaxy has revolved only 20 times since the Sun and solar system formed, 15 times since the earliest microscopic life on Earth existed, 10 times since the Earth's oxygen-rich atmosphere evolved, 5 times since the worms and jellyfish were the most advanced life forms on Earth, and less than one hundredth of a revolution since the appearance of early humans—those curious and perplexing creatures who discovered in the late 1920s, just about the time of the Great Crash, that they were living in a spiral galaxy and that it was revolving every once in a cosmic while.

A Cosmic Turtle's Pace

Our Milky Way Galaxy is flying through the Universe at about 1.4 million miles (2.3 million kilometers) an hour, heading in the direction of the constellation Hydra. No one really knows where we are going, but some scientists believe our 100-billion-plus star spiral is being pulled along by a super-cluster of distant galaxies.

As fast as this galactic velocity seems, it would still take a spaceship over 2,100 years at this speed just to reach the nearest star system, Alpha Centauri, and 50 million years for the Galaxy to cover a distance equal to its own diameter, about 100,000 light-years—a turtle's pace on a cosmic scale.

Traveling Light

Our vast and spiraling Milky Way Galaxy is so large that a powerful flash of light from one edge, traveling at its natural speed of 670 million miles per hour (1.1 billion kilometers per hour), would take 100,000 years to reach the other side.

In Our Galaxy, Far, Far Away

The most distant star in the Galaxy's disk system lies directly opposite us, beyond the dense galactic core and the trillions of miles of obscuring gas and dust. This region of the Galaxy is even closed to the far-probing radio astronomers, who have mapped large areas of the Galaxy by observing neutral hydrogen at the 21-centimeter (8.3-inch) wavelength. Even at one tenth the speed of light, it would take a starship 800,000 years to reach this faraway region of the Galaxy. If the Sun–Earth distance of 93 million miles (150 million kilometers) were represented by 1 inch (2.54 centimeters), the farthest star in the Galaxy from the Sun would be 79,000 miles (127,000 kilometers) away.

Looking Down the Arm

Spiral galaxies, like our own Milky Way, are the most beautiful galactic shapes in the Universe, and it is the spiral arms, trailing away from the galactic motion in varying wisps of cosmic motion and light, that give them their beauty.

The Galaxy's spiral arms are within the disk, where young hot stars are constantly born out of condensing star stuff. This is so because the clouds of gas and dust are thickly distributed along the spiral arms, concentrated there because of the gravitational density waves that flow around the Galaxy. In the 1950s, optical astronomers mapped small areas of three spiral arms in our Galaxy. They did this by tracing the hot bright stars and gaseous nebulae known to populate only the disk. Because of this concentration of gas and dust in the disk, optical astronomers have about a 6,000-light-year limit to their galactic surveying, and only radio astronomers can further plumb the depths.

The Orion arm, our arm, is where the Sun is located, near

the inside lower edge of a local system called Gould's belt, which faces the galactic nucleus about 30,000 light-years away. Some 6,000 light-years farther away from the Center is the Perseus arm, and an equal distance closer, toward the nucleus, is the Sagittarius arm. They were named for well-known constellations located within them.

Next winter, when you see the famous Orion constellation and the sword's gaseous nebula, you will be looking down the Orion arm about 1,600 light-years and seeing an image that left its source in A.D. 380—perhaps to a time before our solar system even entered the arm.

☆ ☆ ☆ ☆ ☆

THE GALACTIC OUTSKIRTS

Relentless Starlight

Besides the Galaxy's spiral disk and dense nucleus, where the majority of stars and intergalactic gas and dust are concentrated, there exists what astronomers call the *galactic halo,* a roughly spherical shape that may extend as far as 500,000 light-years out from the galactic core. This halo contains a sparse population of stars as well as the giant star associations known as globular clusters. About 125 globular clusters are known, all of which move in giant elliptical orbits around the galactic core and define the basic spherical shape of the halo. This shape is generally believed to represent the primitive confines of the Galaxy during its early formation, when it was a vast cloud of gas that would eventually condense into the stars and spiral disk of today.

While the average star population of globular clusters is estimated at 100,000, the larger clusters such as M 13 in Hercules probably contain at least 1 million stars. An average diameter of a globular cluster is 150 light-years.

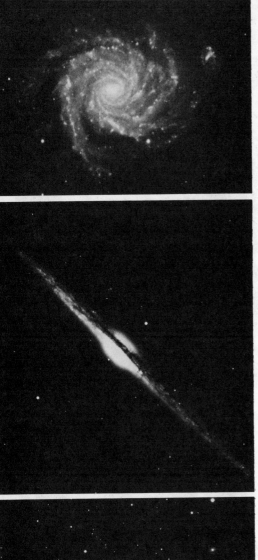

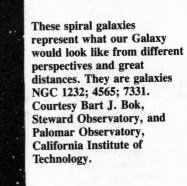

These spiral galaxies
represent what our Galaxy
would look like from different
perspectives and great
distances. They are galaxies
NGC 1232; 4565; 7331.
Courtesy Bart J. Bok,
Steward Observatory, and
Palomar Observatory,
California Institute of
Technology.

In 1974 a message was sent to globular cluster M 13, some 25 thousand light-years away. A reply may come in A.D. 51,974.

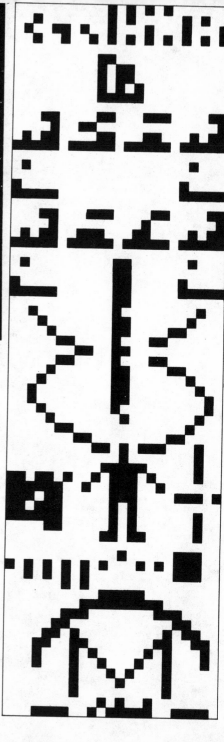

At right, the message. Top line, reading left to right gives numbers 1 to 10 in binary code. Second line: atomic numbers of basic elements. Third line: DNA molecules. Fourth line: double helix of DNA. Fifth line: population, human being, height average. Sixth line: Sun and planets. Seventh line: Arecibo transmitter.

Star densities in globular clusters are among the highest in the Galaxy, holding their own with portions of the dense galactic core. The skies of a planet orbiting a globular cluster star would be brilliant, the stars shining brighter than several full moons. Inhabitants of such a planet would never experience night as we know it, only relentless starlight; nor would the astronomers there be able to see distant cosmic objects, including the great swirl of the galactic disk—home of the Sun, Earth, and other life.

Distant Neighbors

Any good-sized globular cluster containing about 1 million of the Galaxy's oldest stars would have more stars packed in its volume than any other part of the Galaxy with the exception of the galactic core. If each star in the cluster were represented by a golf ball 1 inch (2.54 centimeters) in diameter, then the entire cluster could be contained in a spherical volume with a 10,000-mile (16,000-kilometer) diameter, and the average golf-ball star would still be separated by 100 miles (160 kilometers). Even in the most dense globular cores, the golf-ball stars would have 33 miles (53 kilometers) between them. This would be like your next-door neighbor living about 40,000 miles (64,000 kilometers) away.

The Immense Message

The Earth may receive a message from an advanced galactic civilization in the Great Globular Star Cluster in Hercules. When? In about A.D. 51,974. The message would be in response to the first extraterrestrial radio transmission from the Earth, beamed toward Hercules and M 13 globular cluster in November 1974. The long wait is because of the time it will

take radio signals to make the round trip at the speed of light—
a distance of about 50,000 light-years. Our radio message was
sent by the huge 1,000-foot (305-meter) radio telescope at
Arecibo, Puerto Rico, during the dedication ceremony of the
telescope's new surface. When the message arrives at M 13
about 25,000 years from now, it will have spread out to be as
wide as the entire cluster—150 light-years, a width that is almost 9.5 million times the distance of the Earth from the Sun.
If there are advanced starfolk living somewhere in M 13, they
should hear it.

A Globular Year

The largest globular cluster in the Galaxy, Omega Centauri (NGC 5139), has been known for at least 1,800 years,
but it was believed to be a single star up until the century of the
telescope, when Edmund Halley recognized it as a cluster in
1677. With a diameter of 620 light-years, this globular cluster
may also be the most massive one in our Galaxy, perhaps
equaling 500,000 suns. Its stars exceed 1 million, and their col-
lective light outshines our Sun by 1 million times. Omega
Centauri may also be the oldest globular cluster, with an estimated age of about 13 billion years, dating back to the beginning of the Galaxy.

The orbit, which takes it in the opposite direction of the
spiral disk rotation, swings out 21,000 light-years from the galactic center at its farthest and to within 6,200 light-years at its
closest.

Assuming that the Galaxy is 13 billion years old (a reasonable figure), Omega Centauri has orbited around the Galaxy 130 times since the Galaxy's birth, 50 times since the
Sun's birth, 6 times since hard-bodied animals such as corals
and starfish appeared on the Earth, and 1 time since flowering

Planets in our Galaxy's dense globular clusters would never experience night as we know it; starlight would be too bright. Globular clusters NGC 6522 and 6528. Courtesy Kitt Peak National Observatory; Cerro Tololo Inter-American Observatory.

There is enough gas and dust in the Galaxy's spiral arms to form about 18 billion new stars the size of our Sun. Courtesy Yerkes Observatory.

plants first blossomed on our planet. In just 1 Omega Centauri year, the Earth orbits the Sun 100 million times.

☆　☆　☆　☆　☆
THE SPIRAL DISK

The Transitional 10 Percent

About 10 percent of the Galaxy's mass is composed of interstellar matter—gas and dust from which new stars constantly evolve and which is replenished by dying stars that explode and seed space with their debris. Most of this interstellar gas and dust lies in or near the galactic plane, where it is concentrated in clouds in the galactic arms that trail behind the motion of galactic rotation. About 18 billion new stars the size of our Sun could form from the Galaxy's gas and dust—a total representing one star for each year of the age of the Universe.

The Giant Galactic Clouds

Vast streams of interstellar gas and dust ebb and flow throughout our Galaxy, concentrated mainly in the galactic disk and its spiral arms. Some of the most beautiful cosmic objects seen from Earth are the visible gaseous nebulae such as the Eagle (M 16), the Lagoon (M 8), and the Trifid (M 20), islands of ancient colored light visible only because of young, bright, nearby stars that spotlight them. Without the starlight, the gas and dust clouds are seen negatively, as cosmic holes devoid of stars.

In the late 1970s, radio astronomers discovered evidence

for gigantic, dense molecular clouds that dwarf the bright and dark nebulae, both in dimensions and mass. The nebulae, it turns out, are just small areas that call attention to themselves in the gigantic molecular clouds. Astronomers now believe that a vast molecular cloud flows from the Eagle nebula (M 16) through the Omega nebula (M 17) through the Trifid nebula (M 20) and possibly to the Lagoon nebula (M 8). All these starlit nebulae probably formed out of this colossal cloud, which measures 11 degrees across the sky—a span equal to a train of 22 full moons coupled side by side.

Galactic Smog

The Galaxy's central region is hidden from optical astronomers by 29,000 light-years of intervening gas and dust. To view the galactic center from Earth through all this galactic smog would be analogous to viewing the Sun through clouds so thick that the penetrating sunlight would give off only one thousandth as much light as the full Moon. It would be equivalent to the dimming effect that an ocean depth of several thousand feet would have.

The Dusty Galaxy

The interstellar dust in our Galaxy alone is equal to the mass of 465 trillion Earths.

The Cosmic Grains

More interstellar dust grains can fit into about 12 cubic inches (200 cubic centimeters) than there are stars in the Galaxy—at least 200 billion.

These beautiful starlit clouds of dust and gas, the Eagle, the Lagoon, and the Trifid, are just accents (bright spots) in a gigantic molecular cloud in our Galaxy's disk, with 10 million times more mass than our Sun. Courtesy Palomar Observatory, California Institute of Technology; Kitt Peak National Observatory; Lick Observatory.

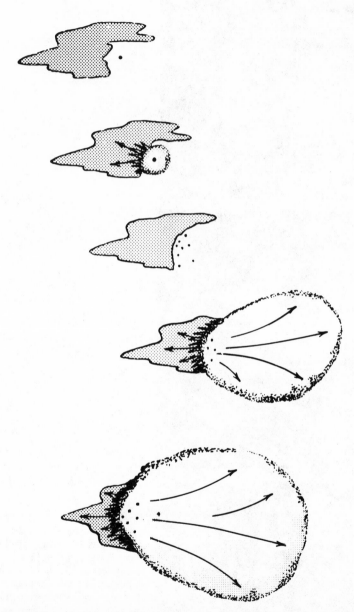

The Cygnus Superbubble was blown by a series of stellar explosions over the last 3 billion years. Total energy involved was 20 times our Sun's output since its birth 5 billion years ago. Courtesy NASA.

Your Everyday Cosmic Molecules

Most of the Galaxy's interstellar gas and dust is confined to a relatively thin layer of the galactic plane in the spiral arms. Technological developments in the 1960s—new satellite and ground-based detectors—have allowed scientists to discover in the last two decades a great many molecules in the "dirty" areas of the Galaxy. These molecular groupings include water, ammonia (household cleaners), formaldehyde (used in wash-and-wear fabrics), carbon monoxide (car exhaust and pollution), and many others, including some rather complex molecules. The most common atoms in these large molecules are carbon, hydrogen, oxygen, and nitrogen—the so-called building blocks of life. Two of the molecules discovered in the Galaxy, methylamine and formic acid, can combine to form glycine and amino acid. Amino acids, of course, combine in great complexity to form the proteins so important for life. That life could actually evolve in the gaseous galactic clouds is an intriguing and controversial subject that has even caught the imagination of novelists. Beware the galactic cloud creature!

A Galactic Cocktail

One of the more interesting molecules discovered in the Galaxy is ethyl alcohol, which was detected in the Sagittarius B2 cloud in the 1970s. This complex molecule of 9 atoms is better known as grain alcohol. Astronomers estimate that this galactic source of grain alcohol, if condensed and purified, could produce 10,000 trillion trillion fifths at 200 proof—an amount of booze that is much greater than the mass of the Earth. But if an astronaut were to fly through such a cloud at the speed of light, holding out a martini glass to catch the alcohol molecules, it would take a century before his glass was full.

The Cygnus Superbubble

About 6,500 light-years away from our solar system, in the constellation Cygnus behind the dark obscuring matter of the Great Rift, is a vast superbubble in the interstellar matter. Discovered by the High Energy Astronomy Observatory satellite from x-ray data in 1980, the supperbubble has a diameter of almost 1,200 light-years, and the rarefied gaseous matter within has a temperature of about 2 million degrees Celsius (3.6 million degrees Fahrenheit).

Astronomers believe that the superbubble was blown by a series of powerful stellar explosions—supernovas—over the last 3 million years. It began when a large star exploded and sent a shock wave into a nearby interstellar cloud of gas and dust. New giant stars then formed in the cloud, lived out their short lives, and in turn exploded. Over time, with more and more supernovae, the bubble inflated. Still more stars formed and exploded, and the bubble expanded to its present size, which is 1/2500th of the Galaxy's entire disk. It is possible that 10 percent of the Galaxy's disk is composed of similar superbubbles that are ideal wombs for stellar gestation.

The energy involved in the Cygnus superbubble is astounding, the most of any single feature in the Galaxy—a whopping 10^{52} ergs, which is up to 20 times the Sun's total energy output since its birth about 5 billion years ago.

Waiting for the Right Planet

The Galaxy's eddies and condensations of gas and dust form an estimated 10 planets and their stars each year, but a habitable planet—one with the right mass, a star of the right mass, and the appropriate distance between them—takes on the average 15,000 years to form, which is the time since Cro-

Magnon Man was producing his cave paintings in the depths of the Ice Age in France.

Galactic Gestation

Every 18 days—about 20 times a year—our Galaxy gives birth to a new star. Every 11,350 years, on the average, a habitable planet forms around one of these stars. Every half second, a human birth occurs on planet Earth.

Counting Planets and Moons

A conservative estimate says there are almost 900,000 habitable planets in our Galaxy (1 such planet for every 227,000 stars) out of a total of some 100 billion planets. If each planet, habitable and nonhabitable, has several moons (a reasonable assumption based on our solar system), there may be as many as 1 trillion moons in the Galaxy. So far, humankind has physically traveled to only one of them.

☆　☆　☆　☆　☆

THE INNER GALAXY

A Galactic-Core Year

The core of our Galaxy, forever hidden to optical astronomers by hot gas and dust, is about 13,000 light-years in diameter, almost one eighth of the full galactic diameter. The image of a sunny-side-up egg with a small yoke approximates the proportions. The core's shape is not spherical, like a globe, but is flattened at the poles and resembles an ellipsoid. This highly concentrated core of our Galaxy accounts for 40 percent of the entire Galaxy's mass and contains as many as 130 billion stars.

The stars and star stuff near the periphery of the core rotate faster than any other region of the Galaxy, except in the extreme central core itself, making 1 revolution every 50 million years. It has rotated more times since the beginning of the Galaxy than any outer galactic region—240 times—so that with 125 more revolutions, taking another 6.25 billion years, 365 galactic-core days will have passed, or 1 galactic-core year. A galactic-core year, then, equals 18.25 billion Earth years—longer than the present age of the Universe.

Star Collisions

Stars probably collide once every 1,000 years or so in the dense stellar orbitways of the galactic core. At this rate, about 12 million stars have collided in the galactic core since the beginning of the Galaxy, about 4.6 million since the birth of the solar system, and 2 since the birth of Jesus at Bethlehem. This is a remarkably high rate when compared to the average time between sun-size star collisions for the Galaxy as a whole, which is 1 every 250,000 trillion (250,000,000,000,000,000) years.

If our automobile collision rate were to match the star collisions in the densest parts of the galactic core, we would have to wait 2 million years for the first auto accident, and there would have been not a single one so far in auto history.

Dizzy Center

The stars, gas, and dust located in the extreme center of the Galaxy, in a region of about 3 cubic light-years, whip

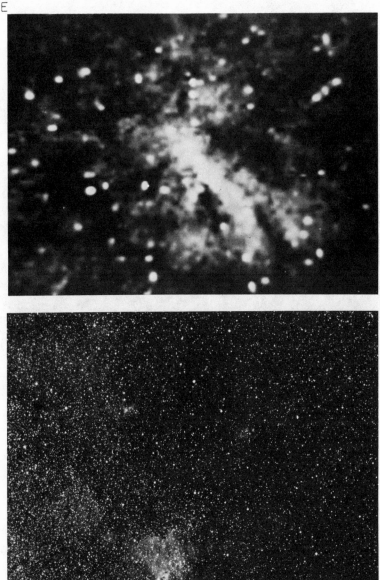

NE

1.1°

Infrared (*top*) and optical (*bottom*) views of the central one-degree of our Galaxy. At infrared frequencies, astronomers can "see" what is visibly hidden behind thick clouds of dust and gas. Infrared power emitted from the center amounts to 1 billion times the Sun's luminosity. Courtesy E. Becklin and G. Neugebauer, Caltech., and *Mercury Magazine*, Sept./Oct. 1979, © 1979 Astronomical Society of the Pacific.

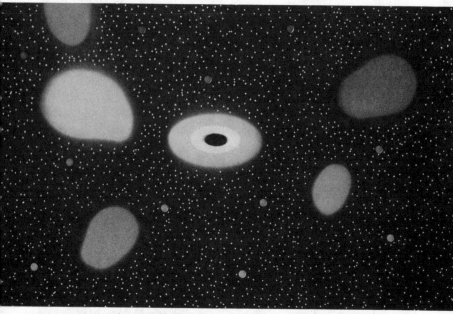

A schematic view of our Galaxy's core. The large patches are dust and gas clouds, the dots are red giant stars. At the center is the black hole and its accretion disk. From "The Central Parsec of the Galaxy" by Thomas R. Geballe. Copyright © 1979 Scientific American, Inc. All rights reserved.

around the center once every 10,000 years. Every time the Sun and solar system revolve once around the Galaxy, the core stars revolve 23,000 times.

Too Close for Comfort?

The highest concentration of stars in the Galaxy is believed to be in a dense star cluster, infrared source number 16, within the radio source region Sagittarius A West, near the very nucleus of the Galaxy. There may be as many as 1 million stars in a space only one fourth of a light-year in diameter, which is equal to about 197 times the diameter of our solar system. If our galactic neighborhood had the same proportionate crowding, there would be another star located at less than 3 times the distance to Pluto, and it would shine on Earth with the brightness of 30 full moons.

Heart of Darkness

The most mysterious object in our Galaxy's central region is a strong and compact radio source with a diameter of 930 million miles (1.5 billion kilometers), about equal to the Earth–Saturn distance, located in Sagittarius A West. At the center of this radio source there may be a supermassive black hole, equal in mass to 5 million suns, with a diameter less than 9 million miles (15 million kilometers), or about 37 times the Earth–Moon distance. Although this size is too small to detect from Earth, the swirling motions of hot gas, indicated by infrared studies, may be the black hole's accretion disk, where infalling matter orbits the gravitational pit before it is captured forever. Such a black hole at our Galaxy's center would capture and devour 1 million Earths' worth of matter every 300,000 years.

THE BIRTH
OF THE UNIVERSE:
THE BIG BANG
BEGINNING

☆　☆　☆　☆　☆
THE FIRST SECOND

The Age of the Universe

The Universe began about 18 billion years ago with a stupendous explosion, often referred to as the "Big Bang." If each of those 18 billion years were equal to a second, the seconds would add up to almost 575 years—20 family generations. The "about 18 billion years" calls for a plus or minus period of 2 billion years, which would put the beginning somewhere between 16 and 20 billion years ago. Scientists predict the Universe has plenty of time left—in which they can calculate a more accurate figure.

The Tiny Universe

Once upon a time, during one of the smallest fractions of a second imaginable after the Big Bang cataclysm, the entire

Universe visible today—from the Earth and the Moon all the way to the farthest galaxies and quasars detectable by present technology—was compressed to a size even smaller than the point of a needle; even smaller, in fact, than an atomic nucleus.

Big Bang Energy

The energy of the Big Bang cataclysm may have equaled the total energy of 10 million billion quasars, each of which equals 300 billion Suns in light energy alone.

A Pinhead's Worth of Universe

If a speck of the one-second-old Universe at 10 billion degrees Celsius could be brought into our solar system—just a pinhead amount with a radius of 0.03937 inch (1 millimeter)—it would equal over 18 times the entire energy output of the Sun! This blazing pinhead could replace the Sun and give the Earth the same amount of energy if it were positioned as far away as Jupiter, over 480 million miles (772 million kilometers) from Earth.

The Beachball Universe

When the Universe was less than one trillionth of a second old, its radius was just over 3 feet (1 meter), the size of a large beach ball, which could be held in a person's arms.

The Infant Universe

Soon after the zero moment of time, less than one billionth of a second after the Big Bang explosion, the Universe was still extremely compressed—its diameter was about 2,000 times the Sun's present diameter and 218,320 times the Earth's diameter. The infant Universe was therefore 1.25 billion miles (2 billion kilometers) in diameter and would easily fit into our present-day solar system, its outer edge falling between the orbits of Saturn and Uranus. If the Earth were represented by a 1-inch-sized spherical cork (2.54 centimeters), the infant Universe would be a 3.5-mile (5.6-kilometer) round lake.

The Cosmic Caldron

At the beginning of the Big Bang, all matter and energy was concentrated at the enormous temperature of 100 billion degrees Celsius! One second after the explosion the temperature had dropped to 10 billion degrees. At about 3 minutes, it had dropped to 1 billion degrees. Seven hundred thousand years later, it was down to 2700 degrees Celsius (4900 degrees Fahrenheit), about 16 times the oven temperature needed to roast a large Thanksgiving turkey.

Before Zero

What happened before the Big Bang? This question is a mystery, and most scientists refuse to even speculate about it. In fact, the question really has no meaning in the context of general relativity, since there was no such thing as space and time *before* the Big Bang. The word *before* itself presupposes the concept of time, which did not exist until the cataclysmic explosion began it all at zero Big Bang.

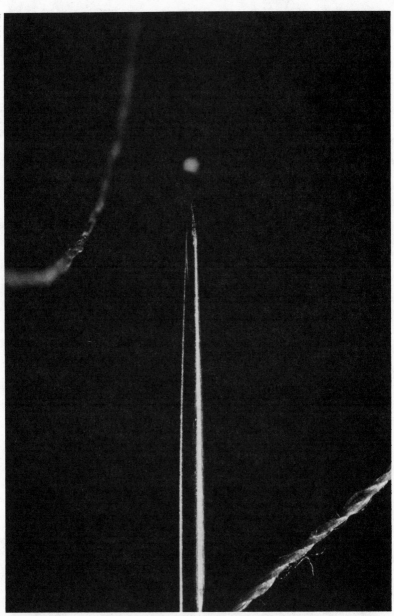
The Universe was once smaller than a point of a needle—sometime before it was a trillionth of a second old. Photo by Doug Nicotera.

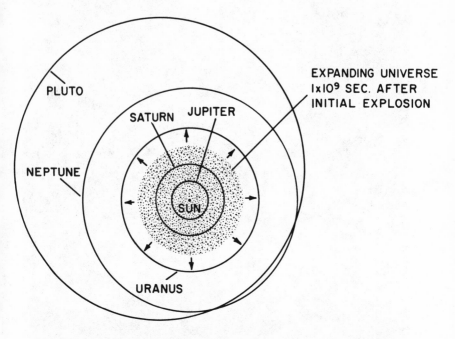

When the Universe was less than 1 billionth of a second old, it was smaller than our solar system. Its edge would have fallen between the orbits of Saturn and Uranus. Courtesy Science Graphics, Copyright © 1982.

The Humbling Singularity

If matter-of-fact-sounding descriptions (theoretical though they be) of the enormous temperatures and densities of the Universe that existed just fractions of a second after the Big Bang cataclysm seem like the ultimate in human presumption, there is always the humbling singularity at Big Bang zero to take the wind right out of the scientists' and theorists' sails. *Singularity* is the strange region in space-time where the laws of physics break down, and the density and temperature of the Universe become infinite. The curvature of space-time also becomes infinite. Even Einstein's General Theory of Relativity cannot describe what happens at singularity. The Big Bang singularity has no doubt caused many restless and sleepless nights for astronomers and physicists.

The Earliest "Time" Ever

Einstein's General Theory of Relativity can predict conditions all the way back to 10^{-43} seconds after Big Bang zero (singularity). Then even this great work breaks down and becomes invalid. This theoretical point in time is known as *Planck time,* named after the German physicist who advanced quantum theory. As a fraction of a second, Planck's instant is

0.0000000000000000000000000000000000000000001
or
One ten-millionth-trillionth-trillionth-trillionth of a second.

Planck time holds the same proportion to the time it takes light to travel 1/360 inch (0.07 millimeters)—the thickness of a thin sheet of paper—as this micro light-time holds to the age of the Universe.

The future of quantum physics will determine our future

knowledge of the earliest fractions of time in the Big Bang Universe. The state of the art of microscopic (quantum) physics is what obscures our view of the Universe before Planck time. A Unified Field Theory will mean a unified theory of the Universe.

The Ultimate Black Hole

Big Bang singularity at Planck time, where infinite temperature and density existed, was the ultimate black hole of the Universe. The primordial stuff of our bodies, part of the Universe, once emerged from this gigantic black hole, and some scientists speculate that our Universe began as a black hole in another universe, which began as a black hole in another universe, which began . . . .

Avoiding the Singularity

If weak gravity did not become strong gravity, in the earliest Planck time Universe (10^{-43} of a second), there is an alternative to singularity. The Hadronic Era (up to 10^{-4} of a second) may have existed for an infinite time before. In fact, developments in radioactivity theory in the 1970s suggested that the force of gravity may not increase its power at Planck time. If this is so, and the Hadronic Era went on before, then there was no beginning. The hadrons (strongly interacting particles such as protons and neutrons) would have gotten heavier, and the earliest Universe would have gotten hotter. The properties of the hadrons would have prevented the singularity—an out-of-this-world point in infinity. Contemplating this idea causes many physicists to perspire.

Before the Beginning

Most Big Bang experts do not feel confident when taking theory back to a time earlier than when the Universe was one hundredth of a second old. Lack of confidence, however, does not stop their speculations. Just before Big Bang zero—one ten-thousandth of a second before—the Universe (if it existed!) was much hotter than 1 trillion degrees Celsius (that is, 1,000,000,000,000; 10^{12}), which is 67 billion times hotter than the core of the Sun. Such a tremendous temperature would tear apart neutrons and protons, reducing them to the basic particles that physicists have postulated: quarks, partons, unitons, and their antiparticles. Where (and when) did the beginning ever end?

Frozen Quantum Theory

One of the more ambitious goals of contemporary physics is to describe the Universe, in the context of the Big Bang model, at a time when it was just one ten-billionth of a second old. This is a rather tall order, considering the fact that most theorists lose their sense of confidence when they approach the first hundredth second of time after Big Bang zero. Nevertheless, elementary particle theory provides some interesting possibilities before the hundredth-of-a-second mark. One of these suggestions is that the extremely early Universe (before one millionth of a second) went through a phase transition, when its temperature fell below a critical point, similar to that reached when water freezes and becomes ice at 0 degrees Celsius (32 degrees Fahrenheit). Before this so-called ''freezing'' occurred, the weak interactions may have been unified with the electromagnetic interactions and a symmetry prevailed in the Universe. After the ''freezing,'' however, the symmetry between the weak and electromagnetic interactions of the elementary particles was broken and has remained so as a characteristic of our Universe. According to this theory of particle physics, the Universe "froze" just at that point in time when its temperature fell below 3,000 million million degrees Celsius. Perhaps our Universe is just a crack in the primordial ice.

When Micro Becomes Macro

Einstein, in his General Theory, described how the relationship between space and time is affected by the gravitational effects of matter. But the General Theory is not able to describe the quantum nature of gravity at Planck time and beyond (10^{-43}), when a temperature of 100 million trillion trillion degrees Celsius was reached. Even expert speculation about particle physics and gravity at such high energies in the earliest microtime of the Universe must be viewed as just that—speculation.

We know that gravity is an extremely weak force relative to the strong and weak nuclear forces and the electromagnetic force, even if it does hold the planets in their orbits around the Sun. Research has not yet discovered any effect that gravity has on the internal subatomic properties of the early Universe; but with the unimaginable temperatures and densities at, or beyond, Planck time, gravity may have become dominant over all other forces, including the strong nuclear force that held sway in the Hadronic Era. If this happened, the extreme gravitational fields would have created particles in large numbers, and each "particle" would have become something entirely different from what it is understood to be today. Each "particle" would have had a "horizon" (a distance-time limit beyond which no light or electromagnetic signal can pass) "closer" than its own wavelength. This is to say that each "particle" would have been about the same size as the entire Universe that can be observed today!

The Hadronic Era

After Big Bang zero (singularity) and the Planck instant (the point before which modern science knows nothing) came the Hadronic Era of the Universe. This entire "era" took place during the first second of the Universe; in fact, it was over when the Universe was one ten-thousandth of a second old. During the Hadronic Era, the Universe continued to expand and at one point was the same size as the Earth.

Hadrons are heavy elementary particles, the simplest forms of matter (neutrons and protons are examples) that participate in strong interactions. The annihilation of proton-anti-proton pairs took place at this time, after which very few hadrons remained. The four basic forces of nature—gravitational, electromagnetic, weak nuclear, and strong nuclear—may have separated out of a primeval force that had sufficiently high energy to combine all forces at the same strength. By the end of the Hadronic Era these forces had changed their relative strengths drastically. Some scientists believe that primordial black holes were created at this time because of slight inhomogeneities in the Big Bang. These black holes may have come in all sorts of sizes, some as small as an atom and some as gigantic as an entire galaxy.

The Big Bang's Tiny Black Holes

Very tiny primordial black holes (mini black holes) may have been created before the Big Bang Universe was one ten-thousandth of a second old, but because of the extreme and unusual effects of quantum physics during this microtime, recent theory says that their properties were entirely different from large black holes. These mini black holes *emit* particles and radiation (most black holes, because of extreme gravitational forces, *capture* all matter so that nothing can escape, not even light), and they evaporate rapidly—so rapidly they explode! A tiny black hole with 100 tons of matter could not exist for more than one ten-thousandth of a second before exploding, but the energy released would amount to over 2 billion tons of TNT or 2,000 one-megaton bombs. This is roughly equal to the total energy of all nuclear warheads in the U.S. arsenal.

A Trifle More Force

Had the strong nuclear force which governs the reactions between nuclear particles been more forceful by a few percent during the first second of the Universe, then helium-2, which does not exist, would have been created in abundance. As a result, the stars would have been formed of helium, not hydrogen, and they would have been unstable, burning quickly, often exploding. Just that trifle more force would have given us a quick Universe.

The Leptonic Era

The Leptonic Era of the Universe began about one ten-thousandth of a second after Big Bang zero, when the temperature had dropped to a modest trillion degrees Celsius. It continued until about the one-second mark, at which time the temperature had dropped to 10 billion degrees. At one second, the size of the Universe was approximately 5.8 light-years (34 trillion miles; 55 trillion kilometers); about the same distance as Barnard's star, the second-closest star to the Sun.

Leptons, unlike hadrons, are light elementary particles that *do not* participate in strong interactions but only in weak interactions. They include electrons, muons, and neutrinos. During the Leptonic Era, the Universe was a mixture of photons, neutrinos, antineutrinos, and (for an extremely brief initial period) electron-positron pairs. First, the neutrinos "decoupled" and began their free expansion outward; this was long before the photons decoupled, hundreds of thousands of years later, which created the cosmic background radiation observed today. After the neutrinos broke loose, the electrons and positrons began to annihilate one another, creating photons in the process. At the end of the Leptonic Era, only one electron survived for every 100 million photons. This conversion of matter to energy brought the Universe to its Radiation Era, which would last for much longer than minute fractions of a second—for about 700,000 years.

The Earliest Neutrinos

When the Universe was much less than a second old, just entering its Leptonic Era, neutrinos "decoupled" and began their free expansion—the first particles to leave the initial thermal equilibrium of the Universe. This event occurred about 700,000 years before radiation and matter decoupled and the photons began their free expansion, which led to the transparency of the Universe—an event fossilized in the cosmic background radiation and detected by radio astronomy.

Theoretically, there is a ''neutrino background'' throughout the Universe because of this early event, but the energy of these neutrinos has decayed tremendously with the expansion of the Universe and the passage of billions of years. Their energy today would only be one thousandth of an electron volt, a billion times weaker than the neutrinos created at the center of the Sun (see Chapter 1, ''The Sun''). Since neutrinos are elementary particles that interact very weakly with matter, they are extremely difficult to detect—even those of the Sun's core that are billions of times more powerful than the Big Bang neutrinos. Big Bang theory has predicted this sea of cosmic background neutrinos at a temperature of 2 degrees above absolute zero Kelvin—that is, −271 degrees Celsius or −455.8 degrees Fahrenheit—which may never be detected. This early neutrino sea is extremely faint, and the technology to detect it is not foreseen. If the neutrino background is ever detected and measured, it will be more important proof of the Big Bang cosmology. There are 100 million neutrinos per atom in the Universe, which does little to confirm the power-in-numbers cliché. This is the same ratio as 100 million blades of grass on a lawn about 100 feet square have to 1 grasshopper jumping about.

The Thick Cosmic Soup

When the Universe was one hundredth of a second old and 100 billion degrees Celsius hot, its radiation was incredibly thick: 4 billion times the density of water you shower in. The difference between this density and lead is in the same proportion as lead has to the vacuum in a TV picture tube.

☆   ☆   ☆   ☆   ☆
THE FIRST MILLION YEARS

The Oldest Light

There was light in the beginning of the universe . . . but the light was trapped in an extremely dense and hot cosmic soup of matter and could not escape for hundreds of thousands of years—the Universe was opaque.

The Three-Minute Recipe

When the infant Universe was three minutes old, it was composed mostly of light, neutrinos (electrically neutral particles), and antineutrinos, with just a sprinkle of nuclear material (hydrogen and helium) and electrons. Several hundred thousand years later, when conditions cooled, the electrons would marry the nuclei and create little hydrogen and helium atoms. These atoms were the start of something big—the Universe as we know it today, filled with galaxies, stars, planets, and people.

Coming in Second

Helium is the second most abundant element—after hydrogen—in the Universe and represents about 25 percent of its mass. It was probably created by nuclear reactions (nucleosynthesis) four minutes after the Big Bang cataclysm, when the Universe had cooled down to 1 billion degrees Celsius.

Helium is found everywhere: in the hot gas surrounding young stars, in the atmospheres of old stars, in cosmic rays, even in quasars. The amount of helium synthesized in the nuclear reaction of stars is extremely small when compared to the cosmic abundance created minutes after Big Bang zero. For all its cosmic abundance, helium is rare on Earth and, in fact, was discovered off Earth in 1868 in the vapors surrounding the Sun. This so-called rare gas, which composes one quarter of the Universe, amounts to only 1/200,000th of the air we breath. A Universe with little or no helium would have affected the condensation of protogalaxies and lowered their density. The result: a Universe without stars, lightless and lifeless.

An Extra Pinch of Helium

If the intense background radiation had been less during the first second of the Universe, the percentage of helium would have been higher—30 percent or more instead of the actual 25 percent abundance—and all the stars in the Universe would burn faster. Our Sun would be dying now instead of having another 5 billion years of stable health to go, and life on Earth would not have had a chance. All this from an extra pinch of helium.

Falling Temperature

In the beginning, the temperature of the Universe fell 99 billion degrees in just 3 minutes—that is a temperature drop of 544 million degrees each second, which is 36 times the temperature of the Sun's core, which in turn is 150,000 times the Celsius temperature for boiling water.

Spread Thin

The infant Universe at 4 minutes was as dense as iron; at 11 minutes, as dense as water; at about 5 hours, as dense as air. Today, some 17 billion years later, life exists in a Universe that, on the average, is spread very thin. We should be thankful that things got a bit clumpy in our local cosmic neighborhood.

The Radiation Era

The Radiation Era began when the Universe was one second old, after the electron-positron annihilation created photons, which then dominated the energy density. At first, gamma rays were abundant, along with neutrinos and antineutrinos. When the Universe was six months old, the temperature dropped below 10 million degrees, and its density also continued to drop. The Universe settled into its infant years, full of a radiant glow, expanding and cooling for the next several hundred thousand years. At midpoint in the Radiation Era, when the Universe was about 350,000 years old, the temperature was 5 million degrees Celsius, as hot as a hydrogen bomb just getting started, when the fireball, which will expand to the size of a large city, is no larger than a homeowner's lot.

The Boring Universe

When the Universe was just over half an hour old (34 minutes 40 seconds), its temperature had fallen to a modest 300 million degrees Celsius, which is 20 times hotter than the core of the Sun. All nuclear processes had stopped, and the Universe continued to expand and cool. For the next 700,000 years, very little happened. Astronomers, had they been alive, would have considered it boring. It was much too early for galaxies, stars, planets, moons, life; in fact, it was still too hot for stable atoms to form. The primordial fireball's glow, thick as fog, was still present at the half-hour mark and would slowly dissipate over the next 700,000 boring years of the Universe—not much longer than the evolutionary span of the starfish on planet Earth.

Happy Birthday, Universe

On the Universe's first birthday, it was still very hot—almost 2.5 million degrees Celsius, which is one sixth the temperature at the center of the Sun and 5,000 times the steak-broiling temperature of an oven. The one-year-old Universe had a "diameter" (scale) of 38,000 light-years, about the same distance that the Sun and Earth are from the Galaxy's center. The density of the Universe on its first birthday was much less than air—somewhere between that of a TV picture tube and the record for a vacuum on Earth.

When's the Matter?

Matter did not exist in the beginning of the Universe; only radiant energy that was 10 million times more dense than matter. At this time there were no atoms, no atomic nuclei—only subatomic particles and antiparticles existed in the radiation: protons, neutrons, electrons, neutrinos. At Big Bang zero, it was not a question of "What's the matter?" but rather "When's the matter?" The matter would have to wait until about 700,000 years after the Big Bang (A.B.B.), when temperatures had dropped to a few thousand degrees and matter began to form from the radiation.

The Decoupling Era

At about the 700,000-year mark, when a threshold temperature was reached, the Radiation Era ended, matter separated (decoupled) from radiation, and hydrogen was created. Hydrogen and helium would begin to form the first galaxies and stars within the next few billion years. It would be another 16 billion years or so before helium-filled balloons began appearing at amusement parks.

A Cooling-off Period

When the infant Universe was less than 1 million years old, perhaps 700,000 years old, it was 2727 degrees Celsius (4941 degrees Fahrenheit). This was the time of decoupling, when matter and radiation separated. Today the cosmic background radiation from this time is −270 degrees Celsius (−454 degrees Fahrenheit). That is a temperature drop of 3000 degrees in about 17 billion years—176 degrees per billion.

☆   ☆   ☆   ☆   ☆

MILLIONS OF YEARS AND COUNTING

Growth Spurt

At the youthful age of 3 million years, the Universe had a radius of 7 million light-years (just 140 times the radius of our Milky Way Galaxy), its temperature was almost 1000 degrees Celsius (1832 degrees Fahrenheit, about 5 times your oven baking temperature), and its density was the same as interstellar space is today—1 hydrogen atom per 0.061 cubic inch (1 cubic centimeter).

The Shirt-sleeve Universe

The Universe was at human body temperature, 37 degrees Celsius (98.6 degrees Fahrenheit), when it was 9.6 million years old and had a radius of 154 million light-years, which is 70 times the distance to the Andromeda Galaxy (M 31), the largest of the nearby galaxies to the Milky Way.

Fast Space

When the Universe was 9.6 million years old and at human body temperature, it was expanding at 1.98 million miles (3.2 million kilometers) a second! Since this velocity is some 10 times faster than the speed of light, this means that space, not matter, was doing the expanding. If it were possible for a spaceship to accelerate this fast, it could travel from Earth to the far planet Pluto in less than 25 minutes.

A Cold Comparison

The coldest temperature ever recorded on Earth is -88 degrees Celsius (-127 degrees Fahrenheit), which occurred at Vostok, Antarctica, August 24, 1960. Our Universe had this average temperature when it was about 21 million years old and hydrogen was probably beginning to condense into the protogalactic clouds.

The Greatest Element

Hydrogen is the greatest element in the Universe. It is the simplest chemical element and consists in its most common

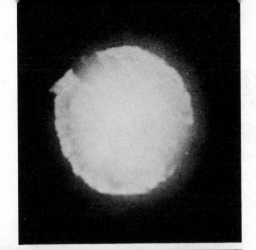

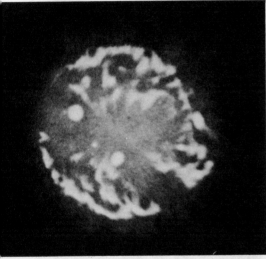

The primeval fireball of the
Big Bang rapidly expanded
and dropped in temperature.
At the 1-second mark, it was
10 billion degrees; at about 3
minutes, it had cooled down
to 1 billion degrees. From the
film, "The Universe,"
courtesy NASA.

When the Universe was 9.6 million years old, it was at human body temperature—98.6 degrees Fahrenheit—a shirt-sleeve Universe. Photo by Doug Nicotera.

form of one proton orbited by one electron; it is by far the most abundant, claiming 73 percent of the mass of the Universe; and it is the lightest, having the lowest density of all elements. One of the few places where it comes in second (in quantity, not importance) is here on Earth, where it combines with oxygen to give us water (H_2O), the basis for all life on our planet. The Earth's crust is made up of only 0.14 percent hydrogen.

Hydrogen formed hundreds of thousands of years after helium, probably when the Universe was not quite 1 million years old. This was the Decoupling Era, when matter and radiation broke out of thermal equilibrium and separated, at a temperature of about 3300 degrees Celsius (5972 degrees Fahrenheit). The relatively low temperatures allowed the electrons and protons to combine to form hydrogen atoms. While the process of hydrogen creation began suddenly, it continued until the Universe had reached its millionth birthday. The Universe today has 1 hydrogen atom for every 295 cubic feet (8.35 cubic meters) of its volume, assuming that all matter from the galaxies and stars was evenly distributed. The vast majority of stars, those on the main sequence, burn hydrogen in their cores and convert it to helium. The energy produced in this fusion process is what has made life possible on the Earth. Our own bodies and those of all animals—indeed, the entire biosphere in which life exists—are composed of 50 percent hydrogen (a percentage of atoms rather than weight). Hydrogen is the blood of the Universe.

Inside the Early Universe

If people could travel back in time and space to observe the early Universe from inside the superhot fireball, what would they see? Some 1,800 years after the Big Bang, they would be bathed in the fireball's radiation, equal to 2.1 billion suns. At the 100,000-year mark, the fireball intensity would have fallen off to equal "only" 210,000 suns. As the temperature dropped further over the next few hundred thousand years, the fireball's glow would go from silvery yellow through the various shades of orange, until the time the Universe approached the Decoupling Era, where radiation and matter separated, when the radiation would be somewhat hotter than molten steel but about the same color. As this 700,000-year-mark approached, the fireball-brilliant space would begin to grow transparent, as if it were a fog thinning out, without fully lifting. This luminous fog, the color and brightness of molten steel, would dissipate slowly, then more markedly, soon allowing a view beyond—blackness, nothing. The surrounding blackness would increase as the hot fog expanded and dissipated further. In time, there would remain a slight glow of yellow or orange, but this too would soon fade, leaving only the blackest space everywhere—no stars, no light, just a hot blackness gradually cooling. Millions of years later, possibly a few billion, condensation would form galaxies and stars, and there would be light once more—light that, 17 billion years later, would enter human eyes and minds and give clues to the beginning of our wondrous and magnificent Universe.

Radiant Dominance

Before the Decoupling Era of the Universe, there was only radiant energy, no atoms of matter, for hundreds of thousands of years. Then there was a short period of density equality between radiation and matter. Today, matter firmly dominates. Humankind, in fact, lives on a cosmic speck of matter, and from this speck called Earth scientists have learned to listen to the early Universe, the cosmic fossil of radiation whose density is now 100,000 times less than the present density of matter over which it once—a long long time ago—held radiant dominance.

The Genesis Fossil

The oldest fossil in the Universe—a fossil of radiation, not of bone—was discovered in 1965 by two Bell Telephone Laboratories scientists, Arno A. Penzias and Robert Wilson. This cosmic fireball radiation, remnants of the Big Bang cataclysm, was found accidentally, when these men were preparing to study microwave emissions from our Galaxy. At first the noise detected by the radio telescope was thought to be extraneous interference, perhaps caused by the atmosphere, close-by objects, or pigeon droppings found inside the antenna! The pigeon droppings were cleaned up, but the noise continued. Soon, with the help of other scientists, Penzias and Wilson established the fact that the noise was the faint, ancient glow of radiation, coming from every direction of the Universe—direct confirmation of the Big Bang theory. This discovery was the greatest and most consequential of modern astronomy, and it has been confirmed in many ways since. The fossil radiation had the predicted wavelength pattern, changed over aeons of time since the Universe began expanding, for the light and heat of the Big Bang explosion—a classic case of tripping over evidence that proves an existing theory (the 1948 work of George Gamow, Ralph Alpher, and Robert Herman).

The temperature of the fossil radiation is about 3 (actually, 2.7) degrees Kelvin (−270 degrees Celsius; −454 degrees Fahrenheit) and has cooled down from 3000 Kelvin degrees (2727 degrees Celsius; 4940 degrees Fahrenheit) when the Universe was 700,000 years old. The cosmic background radiation is said to date back to the beginning of the Universe; "beginning" means this time hundreds of thousands of years after the Big Bang (A.B.B.) when radiation separated from matter ("decoupled"). The opaque Universe suddenly became transparent, and hydrogen atoms began to form. But even though the fossil radiation dates to the first million years and not to Big Bang zero, it still is the oldest signal in the Universe. This echo from creation has come to the Earth from unimaginable durations of time, from farther back in space and time than the most distant galaxies and quasars. It is entirely possible in the next few decades that sophisticated off-Earth astronomy will optically observe this newborn glow of the Universe that shone just after the birth of all there is, more than 17 billion years ago.

The Everywhere Part of the Universe

The cosmic background radiation has been intensely studied and measured by microwave astronomers ever since its discovery in 1965, and the results so far have shown a Universe that is highly uniform in its expansion. One astronomer has called this work "the most accurate measurement ever made in cosmology." This 3-degrees-Kelvin (-270 degrees Celsius) microwave radiation offers a means of measuring the speed of the Earth and the Galaxy relative to the Universe. Slight variations in the intensity of the radiation in different directions of space show astronomers which way the Earth and the Galaxy are moving—that wonderful Doppler effect saves the day once again.* The great spiral Galaxy in which we live, it turns out, is traveling in the direction of the constellation Hydra at 1.4

---

*Doppler effect is the change in apparent wavelength because of the relative motion between the source and observer.

million miles (2.3 million kilometers) per hour. This is faster than expected, and some scientists think that a supercluster of distant galaxies may be pulling us along at this pace. It is hoped that more detailed measurements will eventually lead to evidence for or against some "lumpiness" in the Big Bang beginning. If such unevenness in the background radiation is discovered, it will explain how the galaxies of the Universe were formed. If not, it will be back to head-scratching. The important fact is that the Big Bang has given us a cosmic ruler. Cosmic background radiation is the *only* part of the Universe that is everywhere.

The 1 Percent Factor

The invisible cosmic background radiation, at 2.7 degrees Kelvin (−270 degrees Celsius; −454 degrees Fahrenheit), is the most energetic and oldest signal in the Universe. This cosmic fossil represents 99 percent of all the radiation in the Universe, and the leftover 1 percent includes all the radiation energy from the billions upon billions of stars and galaxies. Life exists because of this 1 percent factor.

The Clumpy Universe

Once hydrogen formation took place in the Decoupling Era, before the million-year mark, the Universe was ready to begin forming great clouds of hydrogen—the protogalaxies—which would then evolve into vast islands of stars and gas such as our Milky Way. This period of protogalaxy formation probably lasted 1 to 2 billion years. What happened during this time, and how an early smooth Universe was changed by gravity into a clumpy one, ripe for galaxy formation, is an intensive area of research, and one of the outstanding problems in understanding the early Universe.

There are many theories, and most share the position that protogalaxies were condensed or accumulated by gravitational forces. One of the more unusual ones suggests that galaxies were formed by matter accreting around primordial black holes (if they existed, of course). Another theory states that quantum fluctuations during the Hadronic Era (the first ten thousandth of a second) were responsible for the eventual gravitational variations that forced the smooth expanding Universe into the clumpy one needed for protogalaxy and galaxy formation. Clumps of matter would have condensed immediately after the Decoupling Era. These early Universe clumps were not your usual-sized ones but amounted to 100 million suns, just enough mass to get small protogalaxies going, some of which then could merge with other small clumps to produce larger clumps—big and small protogalaxies. Add liberal amounts of time, stir constantly, and the eventual result will be us.

SELECTED BIBLIOGRAPHY

Books

Bergamini, David. *The Universe*. 2nd ed. Alexandria, Va.: Time-Life Books, 1977.

Bok, Bart J. and Bok, Priscilla F. *The Milky Way*. 4th ed. rev. Cambridge, Mass.: Harvard University Press, 1974.

Brandt, John C., and Maran, Stephen P., eds. *The New Astronomy and Space Science Reader*. San Francisco: W.H. Freeman, 1977.

Burnham, Robert. *Burnham's Celestial Handbook*. 3 vols. New York: Dover Publications, 1978.

Cloud, Preston. *Cosmos, Earth, and Man*. New Haven: Yale University Press, 1978.

Considine, Douglas M., ed. *Van Nostrand's Scientific Encyclopedia*. 5th ed. New York: Van Nostrand Reinhold, 1976.

Ferris, Timothy. *The Red Limit*. New York: William Morrow, 1977.

French, Bevan M. *The Moon Book*. New York: Penguin Books, 1977.

Hartmann, William K. *Astronomy: The Cosmic Journey*. Belmont, Cal.: Wadsworth Publishing, 1978.

Henbest, Nigel. *The Exploding Universe*. New York: Macmillan, 1979.

Illingworth, Valerie, ed. *The Facts on File Dictionary of Astronomy*. New York: Facts on File, 1979.

Jastrow, Robert. *God and the Astronomers*. New York, W.W. Norton, 1978.

————. *Red Giants and White Dwarfs*. New ed. W.W. Norton, 1979.

————. *Until the Sun Dies*. W.W. Norton, 1977.

Kaufmann, William J. *Galaxies and Quasars*. San Francisco: W.H. Freeman, 1979.

Krogdahl, Wasley S. *The Astronomical Universe*. 2nd ed. New York: Macmillan, 1962.

Maffei, Paolo. *Beyond the Moon*. Cambridge, Mass.: MIT Press, 1978.

————. *Monsters in the Sky*. Cambridge, Mass.: MIT Press, 1980.

Menzel, Donald H. *Our Sun*. Cambridge, Mass.: Harvard University Press, 1959.

Moore, Patrick. *New Guide to the Moon*. New York: W.W. Norton, 1976.

————. *The New Guide to the Stars*. New York: W.W. Norton, 1974.

Motz, Lloyd. *The Universe: Its Beginning and End*. New York: Charles Scribner's Sons, 1975.

Murdin, Paul, and Allen, David. *Catalogue of the Universe*. New York: Crown Publishers, 1979.

NASA. *Images of Mars: The Viking Extended Mission*. Washington, D.C.: National Aeronautics and Space Administration, 1980. NASA SP-444.

————. *Voyager Encounters Jupiter*. Washington, D.C.: National Aeronautics and Space Administration, 1979. JPL 400-24.

Pickering, James S. *1001 Questions Answered About Astronomy*. New York: Dodd, Mead, 1976.

Ridpath, Ian, ed. *The Illustrated Encyclopedia of Astronomy and Space*. Rev. ed. New York: Thomas Y. Crowell, 1979.

Sandage, Allan; Sandage, Mary; and Kristian, Jerome, eds. *Galaxies and the Universe*. Chicago: Unviersity of Chicago Press, 1975.

Sandage, Allan. *The Hubble Atlas of Galaxies.* Washington, D.C.: Carnegie Institution of Washington, 1962.

Scientific American, Inc. *Cosmology + 1: Readings from Scientific American.* San Francisco: W.H. Freeman, 1977.

————. *The Solar System.* San Francisco: W.H. Freeman, 1975.

Shapley, Harlow. *Galaxies.* 3rd ed. Cambridge, Mass.: Harvard University Press, 1972.

Shipman, Harry L. *Black Holes, Quasars, and the Universe.* 2nd ed. Boston: Houghton Mifflin, 1980.

Silk, Joseph. *The Big Bang.* San Francisco: W.H. Freeman, 1980.

Weinberg, Steven. *The First Three Minutes.* New York: Basic Books, 1977.

Periodicals

The following magazines and newsletters were often consulted.

Astronomy. AstroMedia Corporation, Milwaukee, Wisc.

Mercury. The Astronomical Society of the Pacific, San Francisco, Cal.

NASA News. National Aeronautics and Space Administration, Washington, D.C.

National Geographic. National Geographic Society, Washington, D.C.

Science News. Science Service, Washington, D.C.

Scientific American, Scientific American, Inc., New York, N.Y.

Sky and Telescope. Sky Publishing Corporation, Cambridge, Mass.

INDEX